The Microbiome Organ

The Ecosystem Inside Us That Shapes Health

Team Trex

Published by NobleTrex Press

For permissions and other inquiries, write to:

P.O. Box 3132, Framingham, MA 01701,
USA

Contents

Introduction

Look at your hand. Flip it over; examine the unique topographic swirl of your fingerprint, the lifeline cutting across your palm, the sheer *humanity* of the skin. It feels like a boundary, doesn't it? A definitive line where *you* end and the rest of the world begins. You have been taught from your earliest science classes that you are a singular individual-a discrete biological entity driven by the twenty-three pairs of chromosomes inherited from your parents. You are the architect of your own physiology, the captain of a solitary ship.

That lesson was a lie. Or, at the very least, a profound misunderstanding of the biological reality.

If we were to dissolve your body into its constituent cells, sorting them into piles like coins on a table, the pile labeled "human" would be startlingly modest. Depending on when you last used the bathroom, the pile labeled "microbe"-bacteria, archaea, fungi, and viruses-would roughly equal the human stack in sheer number. But if we looked at the genetic catalog, the disparity becomes humiliating. For every one gene in your body that is human, there are approximately one hundred genes belonging to the microbial universe within you.

Genetically, you are less than one percent human. The rest is a bustling, chaotic, ancient, and indispensable metropolis of foreign organisms. You are not a captain of a solitary ship; you are a walking coral reef, a super-organism, a "holobiont."

For centuries, medical science treated this teeming ecosystem as, at best, a collection of hitchhikers, and at worst, a reservoir of potential death. We spent the twentieth century waging a chemical war against the microbial world, armed with antibiotics and antiseptics, operating under the "germ theory" assumption that a sterile body was a healthy body. We were wrong. We now know that by scrubbing away the microbial life, we were not cleaning the house; we were demolishing the foundation.

This book is about the paradigm shift that is currently shaking the bedrock of medicine. We are moving away from viewing microbes solely as pathogens to be eradicated and toward understanding them as a complex, fully functional organ-the Microbiome Organ.

Why call it an organ? It sounds strange to equate a slurry of single-celled organisms in your colon with the structural elegance of a heart or a liver. Yet, consider what an organ does. It is a collection of tissues that perform a specific, vital function for the body. The liver detoxifies chemicals and metabolizes drugs. The microbiome does exactly this. The immune system distinguishes friend from foe. The microbiome teaches it how. The endocrine system regulates metabolism and mood. The microbiome produces the very chemicals that dictate these states. If you were to surgically remove your microbiome-a state scientists create in "germ-free" mice-your body would fundamentally fail. Your heart would shrink, your digestion would falter, your immune system would remain in a state of infantile confusion, and your stress response would go haywire.

We possess a forgotten organ, one weighing roughly as much as the brain, but one that is acquired rather than inherited, fluid rather than fixed.

The implications of this reframing are staggering. If the microbiome is an organ, it can be damaged. It can fail. And, crucially, it can be treated. Many of the chronic, modern plagues that seem to defy simple explanation-obesity, asthma, allergies, autoimmune diseases, perhaps even depression and anxiety-share a common thread: the disruption of this internal ecosystem. We are not just suffering from bad genes or bad luck; we are suffering from organ failure.

Our exploration begins by mapping the architecture of this invisible city. We often speak of "bacteria" as if they are all the same, a monolith of germs. But a healthy gut is a rainforest of diversity. As we will discuss, it matters which species are holding the ground. We have keystone species, akin to the wolves in Yellowstone, whose presence dictates the behavior of the entire ecosystem. We have guilds of workers that specialize in specific tasks, ensuring that the lights stay on and the trash is taken out. When this diversity collapses-often under the bombardment of a Western diet or antibiotic overuse-the resilience of the entire body crumbles.

But how does a human body, which is evolutionarily programmed to attack foreign invaders, tolerate trillions of them living in its basement? This is a question of diplomacy. We will look at the border control mechanisms of the gut, arguably the most complex geopolitical interface on Earth. The gut lining is not a passive tube; it is a sensing station. A single layer of cells, covered in a demilitarized zone of mucus, separates your bloodstream from a world of sewage. Here, a constant exchange of hostages and information takes place. We will dismantle the marketing hype around concepts like "leaky gut" to understand the true science of intestinal permeability, where the breakdown of these borders allows inflammation to seep into the rest of the body, setting fires that can burn for decades.

One of the most poetic aspects of this biology is how we acquire it. You are not born with your microbiome organ fully formed. It is an inheritance, passed down in a messy, beautiful relay race starting at the moment of birth. We will trace this timeline, looking at how the mode of delivery-vaginal birth versus C-section-seeds the landscape of the infant. We will examine the evolutionary brilliance of breast milk. For years, scientists were baffled by the presence of complex sugars in human milk that human infants cannot digest. It seemed like an evolutionary waste of energy. We now know that mothers are not just feeding their babies; they are feeding the specific bacteria that they want to colonize their babies' guts. It is a recruitment strategy, written in sugar.

As we age, this organ matures, stabilizes, and eventually creates a "chemical language" that dictates our well-being. This is where the science becomes truly futuristic. The bacteria in your gut are not just sitting there; they are alchemists. They take the fibers you eat-the ones your own body cannot process-and ferment them into short-chain fatty acids like butyrate. These molecules are not merely waste products; they are fuel for your cells and signals to your DNA. They literally change the way your genes are expressed. We will look at how microbes recycle bile acids and transform amino acids into signaling molecules that enter the blood and travel to distant organs.

This chemical dialogue extends to the most protected fortress in the human body: the brain. The "gut-brain axis" has become a buzzword in health columns, but the reality is stranger than the headlines. Through the Vagus nerve-a superhighway of information connecting the abdomen to the brainstem-and through the release of neurotransmitters, your gut microbes have a direct line to your mood and cognition. We will sift through the evidence separating the solid science from the "psychobiotic" speculation. Can a probiotic treat depression? Does stress reshape your microbiome, or does your microbiome dictate your stress response? The answer, as we will see, is a loop, a bidirectional conversation where the mind influences the gut and the gut influences the mind.

However, this book is not just a catalog of wonders; it is a look at what happens when the relationship turns sour. We are witnessing an epidemic of "ecosystem collapse." We will examine the role of the microbiome in the obesity crisis, asking whether our gut bacteria are becoming too efficient at harvesting energy from our food. We will look at Inflammatory Bowel Disease as a civil war where the immune system loses its tolerance for its microbial neighbors. And we will analyze the terrifying rise of *Clostridioides difficile* infections, a scenario where a single opportunistic pathogen capitalizes on a decimated ecosystem to wreak havoc-a biological cautionary tale of what happens when we scorch the earth with broad-spectrum antibiotics.

So, what can we do? If this is an organ, how do we care for it? This is where we must tread carefully. The wellness industry has pounced on microbiome science with a fervor that often outpaces the data. We will devote a significant portion of this book to "tending the garden," distinguishing between the marketing of probiotics-most of which are mere tourists that pass through your system without unpacking-and the legitimate power of prebiotics and dietary fiber. We will treat diet not as a calorie counter, but as a management plan for a wildlife preserve.

We will also look at the extreme end of intervention: the fecal microbiota transplant (FMT). It sounds repulsive, but the transfer of stool from a healthy donor to a sick patient is the most potent demonstration of the microbiome's power. It is, quite literally, an organ transplant. It creates cures for incurable infections overnight, and researchers are currently testing its limits against conditions ranging from melanoma to malnutrition.

Finally, we must arm ourselves with skepticism. Science is messy, and microbiome science is in its adolescence. We will conclude by learning how to read the blueprints-how scientists actually know what is down there using metagenomics and gnotobiotic models. We will discuss the "Bubble Boy" mice-animals raised in sterile plastic bubbles-that have provided the strongest proof of cause and effect, proving that microbes drive physiology, not just correlate with it. But we will also learn why a mouse is not a human, and why a headline claiming a cure for Alzheimer's in a rodent model does not mean you should rush to the supplement aisle.

The goal of these chapters is not to give you a prescription or a rigid diet plan. It is to change your self-image. For too long, we have viewed health as a war against nature, a fortress to be defended. But if you are a holobiont, if you are a composite organism, then the war is a civil war. The path to health is not sterility; it is husbandry. It is the careful tending of the complex, invisible, and vital ecosystem that makes you, you.

The silence of the room you are sitting in is an illusion. You are a bustling crowd. You are a hive. Let's meet the rest of you.

Chapter 1

The Invisible Organ

We tend to view ourselves as biologically autonomous entities, bounded strictly by our skin. Yet, the metabolic and immune capacities essential to our survival are partly outsourced to trillions of non-human cells residing within us. This chapter dismantles the view of microbes as mere passengers, revealing a fully integrated 'distributed organ' that is as vital to our physiology as the liver or the heart.

1.1 More Than a Passenger

For most of modern medical history, the microbes living in and on the human body were treated like suspicious strangers. If a bacterium appeared in the bloodstream, it was an emergency. If it appeared in a wound, it was an infection. If it appeared in a culture dish from the gut, it was often dismissed as "normal flora"—a phrase that sounded reassuring precisely because it was vague. "Normal" implied harmlessness; "flora" implied greenery, something decorative and passive. The underlying assumption was simple: the human body was the main character, and microbes were background extras. Useful, perhaps, in the way a well-behaved passenger is useful on a long drive: they do not interfere.

That picture has become hard to defend. Not because microbes have suddenly become more important, but because we have become better at noticing what they have been doing all along.

A turning point came when biologists began to look at microbial life not as a list of species but as a set of chemical capabilities. A classic anatomical organ is easy to define because it has boundaries. The liver is a lump of tissue, with blood flowing in and out through named vessels, and with cell types you can recognize under a microscope. The microbial community in the gut has no such clean edge. It is dispersed; it changes; it is partly attached to mucus, partly free in the lumen; it is acquired rather than built from a single embryo. Yet it performs tasks that are recognizably organ-like: transforming chemicals, producing biologically active molecules, and shaping the conditions under which the rest of the body develops and maintains itself.

Calling the microbiome an "organ" is not a poetic trick. It is a claim about physiology. An organ is not defined by whether it is made of human cells; it is defined by whether the body has come to depend on it for essential functions. By that standard, the gut microbiome earns its title the way an experienced staff member earns theirs: you do not notice them most days, but the moment they are gone, the system falters in ways you can measure.

One reason this shift took so long is that the microbial world is inconveniently invisible to our habitual tools. Traditional microbiology relied on growing organisms in culture, but most gut microbes do not grow easily on standard laboratory media. In the twentieth century, this created a powerful bias: the organisms we could name and study were the ones willing to live on a plate. In the gut, however, many species are specialists that have adapted to an environment with low oxygen, abundant mucus, complex fibers, and constant competition. They thrive in the intestine and sulk in a petri dish.

When DNA sequencing made it possible to survey microbial communities without culturing them, the gut suddenly looked less like a sparse garden and more like a metropolis. In 2010, a landmark metagenomic effort led by Junjie Qin and colleagues assembled a catalogue of microbial genes from human stool samples and reported a gene set vastly larger than our own. The headline figure—often paraphrased as "we have far more microbial genes than human genes"—was not meant to be numerology. It was a statement about biochemical reach. Genes encode enzymes, transporters, and molecular machines. A large collective gene repertoire means a large collective ability to break down, synthesize, detoxify, and signal. The human genome is remarkably stable; the microbial gene pool is expansive and adaptable. Together they form a combined system with capabilities neither side possesses alone.

This is where the idea of the *holobiont* becomes useful. The term refers to a host and its associated microbes considered as a single ecological and evolutionary unit. It does not mean the microbes are genetically fused with us, or that we lose our individuality. It means that, in practice, our bodies have been built and tuned in a world where microbial partners are expected. A fish expects water; a human expects microbes. Remove the expected environment and the organism behaves strangely.

A vivid demonstration comes from germ-free animals: mice raised from birth in sterile isolators, without any microbes at all. They are not simply "cleaner" versions of ordinary mice. They have smaller immune structures, altered gut anatomy, different metabolic profiles, and changes in the way their intestinal lining turns over. Their immune systems are not merely unchallenged; they are undertrained. The point is not that germs are good in some sentimental sense. The point is that the immune system is a learning system, and one of its teachers is the microbial community that arrives early, stays close, and provides a constant stream of molecular cues. Without those cues, a fundamental part of mammalian development runs in an abnormal mode.

If this sounds abstract, consider a more human-scale anecdote from the early days of microbiology. In the late nineteenth and early twentieth centuries, physicians debated whether the intestine was inherently toxic. Some feared "autointoxication": the idea that putrefaction in the colon produced poisons that seeped into the body and caused illness. This anxiety fueled drastic interventions, including extreme diets and, in some cases, surgical removal of parts of the colon. With hindsight, many of these fears were overstated and often misguided. Yet the premise—that gut microbes can manufacture molecules with systemic effects—was not entirely wrong. The mistake was to assume the effects were necessarily poisonous. The deeper truth is more interesting: microbes are chemists, and the gut is a chemical factory whose products can influence organs far from the intestine.

The liver provides a useful comparison because it is the body's flagship chemical-processing organ. It modifies drugs, neutralizes toxins, reshapes hormones, and packages nutrients. But the liver does not work alone. Many compounds that reach it have already been altered by microbial enzymes upstream. Some drugs are activated or inactivated by microbes before absorption. Some dietary components are transformed into metabolites that the liver then handles as if they were made by our own tissues. Even bile acids, synthesized in the liver to emulsify fats, are not finished products until gut microbes modify them. In other words, the liver is not simply a standalone processor; it is part of a two-stage system in which microbial chemistry often performs the first pass.

This distributed arrangement helps explain why the microbiome can be both resilient and disruptive. When it works well, it expands the range of usable foods and buffers us against environmental chemicals. When it is disturbed—by antibiotics, drastic diet changes, illness, or other stressors—the same chemical power can generate a different set of outputs. The organ analogy does not guarantee benevolence; it guarantees influence. The pancreas is an organ, and pancreatitis is a disaster. The heart is an organ, and arrhythmias kill. A microbiome can be a vital contributor to health and also a source of pathology when its functions shift.

Vitamins provide a straightforward example of mutual dependence. Humans cannot synthesize several vitamins and must acquire them from diet. Gut microbes contribute to this economy by producing certain B vitamins and vitamin K. For a person eating a varied diet, microbial contribution may feel like a bonus rather than a necessity. But in the logic of physiology, redundancy is not trivial. The body often relies on multiple inputs to secure a crucial output. A back-up generator still counts as part of the power system when the electricity goes out. Microbial vitamin synthesis is one of the ways biology hedges against fluctuating food supply and variable absorption.

Fiber is an even better illustration because it exposes a blunt limitation of human anatomy. Many plant carbohydrates are indigestible by human enzymes. We lack the molecular tools to cut specific chemical bonds in complex polysaccharides. If food were only about calories, fiber would be pointless: it would pass through like packing material. But the gut microbiome treats fiber as a resource, fermenting it into short-chain fatty acids such as acetate, propionate, and butyrate. These molecules are not waste. They are absorbed by the host and used as energy sources and signals. Butyrate, in particular, is an important fuel for colon cells and is linked to the maintenance of the gut barrier. Here the "organ" idea becomes concrete: a major nutritional pathway depends on microbial metabolism. The food is not fully food until the microbes have had their turn.

This also reframes a familiar modern puzzle. Many people in industrialized societies consume diets low in diverse fibers and high in refined carbohydrates and fats. The body can digest the refined parts without microbial help, but the microbial community is then starved of the substrates that support some of its most beneficial outputs. In ecological terms, the host changes the habitat and the supply chain; the microbes respond by shifting their population and their chemical production. In physiological terms, the host has altered the function of an organ by changing its inputs. The microbiome is not a passenger. It is a partner whose behavior is shaped by what we feed it.

A skeptic might object that we survive without thinking about any of this. Most people do not collapse into illness after a short course of antibiotics, and plenty of individuals with unremarkable diets live long lives. That is true, and it is exactly why the organ analogy needs to be handled carefully. The microbiome is not a single gland with a single output; it is an ecosystem with overlapping functions. Its contribution is often felt most clearly under stress: during infection, after antibiotics, during early development, in the context of chronic inflammation, or when diet and lifestyle push the system toward a less resilient state. A seatbelt is easy to ignore until you need it.

The clearest clinical example of how essential microbial functions can be is *Clostridioides difficile* infection. This bacterium can lurk harmlessly or be absent entirely, but after antibiotics disrupt the gut community, it can bloom, produce toxins, and cause severe diarrhea and colitis. The striking feature is not only the pathogen; it is the loss of what microbiologists call *colonization resistance*: the ability of a healthy microbial ecosystem to prevent invaders from taking over. This resistance is partly direct competition for nutrients and space, partly the production of inhibitory compounds, and partly the way resident microbes stimulate the host to maintain a defensive posture in the gut lining. In this story, the microbiome behaves like an immune organ as much as a metabolic one. Antibiotics do not merely kill microbes; they can temporarily remove a layer of defense that the body has come to rely on.

One of the most dramatic therapies to emerge from this understanding is fecal microbiota transplantation (FMT), in which stool from a healthy donor is introduced into a patient's gut to restore microbial community function. The idea can sound medieval, but its success in recurrent *C. difficile* infection has been hard to ignore. FMT is not simply adding a probiotic strain; it is rebuilding an ecosystem. In organ terms, it is closer to a transplant than to a supplement. The fact that such a procedure can rescue patients who fail standard treatments is a blunt argument that the microbiome is not optional.

Still, a nagging question remains: if the microbiome is an organ, where are its blueprints? The liver develops from embryonic tissues following genetic programs. The microbiome is acquired. Newborns receive microbes from their mothers, caregivers, milk, skin contact, and the environment. Over time, the community becomes more complex and stable, shaped by diet, infections, antibiotics, geography, and countless small exposures. This difference—inherited tissue versus acquired community—is precisely what makes the microbiome an "invisible organ." It is built by ecological assembly rather than by embryology.

That does not make it less real. Consider another organ-like system we acquire: the immune repertoire. The basic architecture of the immune system is inherited, but the specific antibodies and immune memory each person carries are shaped by exposure. Two genetically identical individuals can have different immunological histories. Yet no one doubts the immune system is a core component of the body. The microbiome is similar: a host-provided scaffold and set of rules, filled in by experience.

Thinking in terms of the holobiont also changes how we interpret individuality. If you ask, "What is a normal microbiome?" you might expect a checklist of species. Yet large surveys, including the Human Microbiome Project's early work, found that healthy people can differ widely in which microbial species they carry. The surprise was that functions can be more stable than membership. One person's gut may contain microbial species A, another species B, yet both ecosystems may carry genes for the same metabolic pathways: breaking down certain fibers, synthesizing particular vitamins, metabolizing bile acids, resisting pathogens. Ecology provides a useful concept here: functional convergence. Different communities can do the same jobs, much as different cities can provide food, sanitation, and transportation with different layouts and populations.

This is why the microbiome-as-organ idea does not require a fixed cast of characters. A liver is made of liver cells. A microbiome organ is made of *functions* spread across many possible microbial lineages. The organismal boundary is less about anatomy and more about metabolic integration. The host supplies habitat (warmth, water, mucus, nutrients, controlled oxygen levels). The microbes supply biochemical transformations and a steady stream of molecules that influence host cells. Between them sits the gut lining: a living interface that is neither mere wall nor passive filter. It is an active mediator, sensing microbial products, adjusting mucus secretion, producing antimicrobial peptides, and deciding which signals count as routine and which count as alarm.

The practical payoff of this view is subtle but profound. If microbes are passengers, then the best medicine is eradication whenever they misbehave. If microbes are an organ, then the best medicine often looks like management: supporting beneficial functions, restoring balance after disruptions, and changing inputs so that the ecosystem produces healthier outputs. This does not mean romanticizing bacteria or treating every microbe as a friend. It means taking seriously that human physiology has been built to operate with microbial partners, and that many modern interventions—antibiotics, processed diets, highly sanitized environments, certain drugs—act on us partly by acting on them.

At a dinner table, it is easy to imagine the liver as a diligent worker in the background: quietly processing a glass of wine, packaging nutrients from a meal, keeping blood chemistry within safe bounds. It is harder to imagine the gut microbiome doing something similarly purposeful, because it does not look like an organ, and because its members have names that sound like tongue twisters rather than like parts of the body. Yet the logic is the same. An organ is a set of coordinated capabilities that the body expects and depends upon. The microbiome earns that status not by being tidy, but by being indispensable.

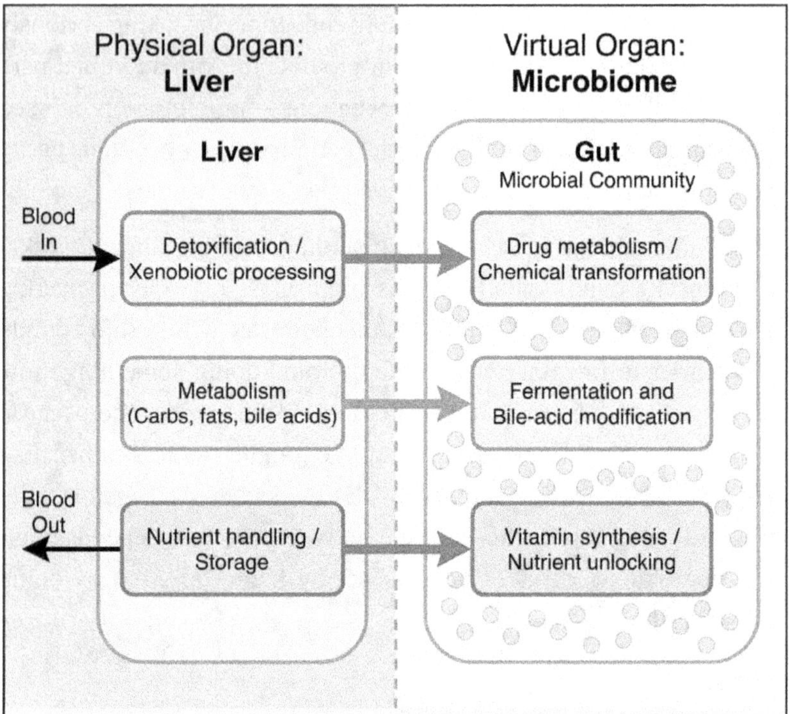

Split-view comparison: A bounded organ (liver) and a distributed "virtual organ" (microbiome) perform functionally parallel chemical work.

If the word *organ* still feels like a stretch, one last thought helps. A person can survive with part of their liver removed because the remaining tissue compensates; the body is full of spare capacity. Dependence does not always look like fragility. It can look like a quiet assumption built into design. Mammals evolved with guts that invite microbes, feed them, tolerate them, and listen to them. The result is not a body plus an infestation, but a composite physiology: human cells and microbial cells cooperating, competing, and negotiating in a shared space, producing the chemistry that makes ordinary living possible.

1.2 The Stability of Chaos

A tidy mind likes tidy systems. If the microbiome matters so much, the temptation is to imagine a healthy gut as a well-disciplined team: the same members showing up every day, each keeping to their station, none causing trouble. The reality looks messier. Two healthy people can share a home, a diet, even a bed, and still carry gut communities that resemble different ecosystems. One person may be rich in microbes that specialize in digesting plant fibers; another may host a different set of specialists and yet remain perfectly well. Even within the same person, the microbiome changes with travel, stress, illness, medication, and the banal fact that Tuesday's meals are not Monday's. If health required a single ideal lineup, most of us would be unwell most of the time.

The surprise—and for many researchers, the relief—is that biological stability does not necessarily mean biological sameness. In ecology, a forest can remain a forest even if no one tree is guaranteed to live long. In economics, a currency can remain stable even as individual prices move daily. In the gut, stability often comes not from a fixed cast of microbial species but from *functional redundancy*: the presence of multiple different organisms capable of doing similar jobs. When the job matters more than the identity of the worker, diversity becomes a form of insurance.

Functional redundancy sounds like bureaucratic waste until something goes wrong. It is the reason commercial aircraft have multiple redundant systems and why the body has paired kidneys. Redundancy creates resilience: if one component fails, another can compensate before the whole system tips into crisis. The microbiome is full of such overlapping capabilities because it is not a machine designed by an engineer; it is an ecosystem assembled by evolution and shaped by daily life. Many microbial lineages have independently evolved ways to harvest energy from the same classes of molecules or to tolerate the same environmental stresses. That overlap is precisely what allows a gut community to keep producing broadly similar outputs even when the membership shifts.

A concrete example begins at the dinner table with a bowl of oats. To the human digestive system, a large fraction of the carbohydrate content in oats is not directly accessible. Complex fibers arrive in the colon largely intact, and there a whole network of microbes begins to dismantle them. Several unrelated bacterial groups can break down parts of these fibers, often using enzyme toolkits that overlap in function even if they differ in details. If one fiber-degrading species is reduced—by a change in diet, an infection, or a course of antibiotics—another may step in, not because the ecosystem is benevolent, but because the ecological niche is profitable. The host experiences this as continuity: fermentation continues, short-chain fatty acids continue to be produced, and the gut lining continues to receive familiar signals.

That continuity is not guaranteed. It depends on the ecosystem having enough diversity to contain substitutes. Low diversity is like a factory that relies on a single technician who knows how to run a critical machine. As long as that person stays healthy, the factory looks efficient. When they take sick leave, production halts.

This brings a different kind of order into focus. The microbiome's apparent chaos—the shifting relative abundances, the person-to-person variation, the long list of Latin names—can mask stable *functions*. Many studies comparing healthy individuals have found that while the specific microbial species present can differ dramatically, the collective set of metabolic pathways inferred from microbial genes can be more consistent. This is not magic and it is not perfect. It is the predictable result of redundancy: multiple community compositions can converge on similar outputs.

The word *dysbiosis* is often used to describe an unhealthy microbiome state, but it can mislead because it sounds like a tidy diagnosis: a single bad arrangement that needs correcting. In reality, dysbiosis is closer to a failure mode. It can mean loss of diversity, loss of key functions, overgrowth of organisms that thrive in inflammation, collapse of colonization resistance, or a community that produces a different balance of metabolites. Different diseases may involve different versions of dysbiosis, and the same disease label may encompass multiple microbial patterns. The unifying idea is that the ecosystem has lost some of its buffering capacity. When shocks arrive, it does not absorb them; it amplifies them.

Antibiotics provide the clearest example of how a buffering system can fail, because antibiotics are an intentional ecological shock. A broad-spectrum antibiotic is not a sniper; it is a wildfire. It may remove the pathogen that prompted the prescription, but it also kills bystanders, some of which were carrying out functions that the host had quietly outsourced. The gut community can recover after antibiotics, but recovery is not always a simple return to the previous state. It can be a reshuffling. Some organisms may return quickly; others may not return at all. If recovery occurs in an environment that has changed—a different diet, ongoing inflammation, hospitalization exposure, repeated antibiotic courses—the community may settle into a different configuration.

The story of *Clostridioides difficile* infection captures this in a dramatic, almost theatrical way. *C. difficile* produces hardy spores that can survive in hospitals and long-term care facilities. Many people encounter the organism without trouble. The trouble begins when antibiotics remove enough of the resident community that *C. difficile* can germinate, expand, and dominate. The toxins it produces damage the gut lining, leading to severe diarrhea and colitis. From an ecological perspective, *C. difficile* is not merely an invader; it is an opportunist that exploits an ecosystem with reduced diversity and reduced competition. From a physiological perspective, this is what it looks like when colonization resistance fails.

The striking part is how often recurrent *C. difficile* infection can be halted by restoring community complexity. Fecal microbiota transplantation (FMT) works in many cases not by targeting *C. difficile* directly, but by rebuilding an ecosystem dense enough to keep the pathogen in check. The practical message is that diversity itself can be therapeutic. The deeper message is that microbial ecosystems behave like other ecosystems: once they have been simplified, they can become easier to invade.

If antibiotics show how diversity can be lost, diet shows how diversity can be reshaped, sometimes remarkably quickly. A modern experimental highlight came from work by Lawrence David and colleagues (2014), who demonstrated that extreme short-term dietary shifts in humans—for example, moving to an animal-based diet versus a plant-based diet—can alter gut microbial community structure within about a day. The detail matters because it undermines a comforting myth that the microbiome is sluggish, like a geological feature that changes only over years. Some parts of it are stable, but much of it is responsive and opportunistic. Feed a new substrate and organisms that can use it expand. Remove a substrate and their competitors may take over.

This responsiveness is a source of both anxiety and hope. Anxiety, because it suggests the microbiome can be nudged into less helpful states by modern eating patterns. Hope, because it implies that at least some microbial functions are tractable: change inputs and outputs can change. Yet the response is not infinitely flexible. The set of possible responses is constrained by which organisms and genes are present to begin with. A garden cannot grow what it has no seeds for. Diversity becomes the library of options available for adaptation.

The ecological language can feel detached until it is tied to a bodily sensation: the difference between a gut that feels robust and a gut that feels easily upset. Some people seem able to tolerate travel, dietary indulgence, and minor infections with little gastrointestinal drama; others develop prolonged symptoms after what looks like a mild perturbation. Individual experience is not a clean experiment, and many factors beyond microbes are involved, but the notion of resilience is intuitive. A resilient system returns to baseline. A fragile system tips into a new state and stays there.

Ecologists talk about alternative stable states: the idea that an ecosystem can settle into different configurations that are each self-maintaining. A clear lake can become an algae-dominated lake and remain that way even after the original trigger is removed, because the new state alters the environment in ways that favor itself. The gut may have analogous dynamics. One influential proposal in microbiome science was the enterotype hypothesis, which suggested that human gut communities cluster into a small number of distinct compositional types. The enterotype idea was valuable because it made researchers ask whether there are a few common attractors rather than a smooth continuum. It also became contentious, because later analyses often found gradients rather than discrete bins, and because clustering can be sensitive to methods and cohorts. Still, the underlying ecological question remains: are there microbiome configurations that, once established, tend to reinforce themselves?

Functional redundancy complicates this question in a productive way. If multiple community compositions can perform similar functions, then even a shift in composition does not necessarily imply a shift in physiological state. A person might move along a compositional gradient and remain functionally stable. Conversely, a community might look superficially similar by taxonomy yet lose critical functions if key metabolic capacities are diminished. This is why microbiome research increasingly emphasizes *what the community is doing* rather than *who is there*.

For a concrete illustration, return to fermentation. The production of butyrate in the colon is often associated with gut health because butyrate is used as fuel by colon cells and is linked to barrier maintenance and immune signaling. Many different microbes can contribute to pathways that ultimately lead to butyrate production, sometimes directly and sometimes through cooperation. In a diverse community, the butyrate "pipeline" can have multiple routes. One organism may be the primary fermenter of a fiber into intermediate products; another may convert those intermediates into butyrate. If one species falters, another may keep the pipeline functioning. In a simplified community, the pipeline may break with the loss of a single link.

This kind of redundancy is not only about metabolism; it is also about defense. A diverse microbiome creates many overlapping obstacles to invasion. Some residents consume resources so efficiently that little remains for a newcomer. Others produce bacteriocins or acids that inhibit growth. Others stimulate the host to produce mucus and antimicrobial peptides, shaping the habitat in a way that favors residents. No single mechanism is universal; the point is that a diverse community can deploy many partial defenses at once. Pathogens then face a multilayered barrier rather than a single locked door.

The most revealing evidence for predictability in this apparent chaos comes from controlled microbial communities in gnotobiotic animals. When researchers introduce a defined set of microbial species into germ-free mice, they can observe how the community responds to specific dietary ingredients under controlled conditions. Work by Jeremiah Faith and colleagues (2011), using a defined ten-member human gut community in germ-free mice, showed that some changes in species abundance can be modeled from diet composition and community membership. This does not mean the microbiome is a simple equation. It means it is not pure whim, either. The ecosystem follows rules: resource availability, competitive interactions, and metabolic dependencies. Diversity does not erase those rules; it provides multiple ways for the rules to play out.

There is a subtle but important corollary. Diversity is not automatically good in every sense. A diverse ecosystem can contain organisms capable of making harmful metabolites under certain conditions, just as a diverse economy contains actors capable of fraud. What tends to correlate with health is not diversity as a moral virtue but diversity as a stabilizing feature that keeps key functions within a healthy range. When the gut lining is intact, immune signaling is balanced, and diet supplies substrates that favor beneficial fermentation, redundancy and diversity often translate into resilience. When the host environment shifts toward inflammation, high oxygen leakage into the gut lumen, or repeated antimicrobial pressure, the same principles of ecology can favor different winners: organisms that thrive in stress and that may further intensify stress.

This is one reason microbiome science has had to develop a new kind of caution about causation. It is tempting to look at a disease and ask, "Which microbe caused it?" Sometimes a clear answer exists, as with *C. difficile*. More often, the microbiome's role looks like a web of small contributions: shifts in metabolites, barrier function, immune tone, and pathogen resistance. Diversity might be lower in a disease state, but that could be cause, consequence, or both. Chronic inflammation can reduce microbial diversity by changing the gut environment; reduced diversity can then reduce resilience and make inflammation harder to resolve. The loop does not offer a single villain; it offers dynamics.

At the human scale, these dynamics show up in an almost paradoxical way. People often search for the "right" probiotic or the "one" missing bacterium. That desire makes sense: it is easier to imagine repairing an organ by replacing a missing part. But in an ecosystem, missing parts may be less important than missing *relationships*. Functional redundancy means no single species is always essential; it also means that what matters is whether the community can carry out certain transformations reliably. Sometimes that reliability can be restored by introducing particular microbes. Sometimes it requires changing the environment so that existing microbes can do their jobs. Often it requires both.

A resilient gut microbiome can be thought of as a crowd that keeps a city running. When a subway line fails, buses reroute. When one grocery store closes, others pick up demand. People complain, life is inconvenient for a while, but basic services continue. A fragile microbiome looks more like a company town: one employer leaves and everything collapses. The seeming disorder of a diverse microbial ecosystem is the source of its quiet stability. In the gut, as in other complex systems, a little chaos can be the price of robustness, and a too-orderly simplicity can be a prelude to breakdown.

1.3 Architects and Workers

At a certain point in any conversation about the microbiome, the names take over. *Bacteroides*, *Ruminococcus*, *Akkermansia*, *Faecalibacterium*. The lists can feel like birdwatching with an unfamiliar field guide: impressive, but not immediately illuminating. Taxonomy matters, and there are moments when a single species behaves like a culprit or a savior. More often, though, the names distract from a simpler question that humans instinctively understand: what is getting done?

A city does not function because it contains a particular person named Smith. It functions because there are people doing certain jobs: running power stations, collecting trash, baking bread, fixing pipes, keeping order. Individual workers come and go; the city survives if the roles are filled. The gut microbiome is like that. It is an economy of chemical work. The organisms may vary between healthy individuals, but many of the essential jobs remain remarkably consistent: breaking down complex carbohydrates, shaping the mucus barrier, converting bile acids, producing short-chain fatty acids, and resisting invaders. Thinking in terms of *guilds* and *keystone species* is a way of taking the microbiome seriously as an ecosystem rather than a stamp collection.

A guild, in ecology, is a group of organisms that use similar resources or perform similar functions, even if they are not closely related. The guild idea is intuitive in the gut because the work is chemical. If two bacteria both ferment resistant starch into short-chain fatty acids, they belong to a functional guild whether or not they share a recent ancestor. This framing has a practical advantage: it helps explain why people can have different microbial membership and yet maintain similar physiological outcomes. It also helps explain why certain disturbances are dangerous. When you lose a few species but keep the guild, the ecosystem can keep functioning. When you lose the guild, the output changes, sometimes abruptly.

A keystone species is different. In architecture, a keystone is the wedge-shaped stone at the top of an arch; remove it and the arch collapses. In ecology, a keystone species is not necessarily abundant, but it has an outsized influence on the structure of the community. The classic examples are not microbes at all: sea otters controlling sea urchins, wolves reshaping deer behavior. In the gut, keystones are often organisms that create or maintain habitats, unlocking resources for others or shaping the interface between microbes and the host. Their importance lies not in the molecules they personally produce but in the way they allow other organisms to thrive.

The distinction between architects and workers makes these roles easier to visualize. *Architects* shape the environment: they build, remodel, and police the spaces where other microbes live. *Workers* process inputs into outputs that matter to the host: metabolites, vitamins, immune-modulating signals. These are not moral categories, and a single organism can play both roles, but the division captures a fundamental truth about ecosystems: structure and function are entwined, and sometimes the most influential organisms are those that control access.

A useful place to see this logic in action is the thin, slippery layer of mucus that coats the intestinal wall. The gut is not a simple tube through which food passes. It is a highly regulated interface, an inhabited border. The host produces mucus to protect the epithelium and to manage contact with microbes. Many microbes, in turn, live in or near that mucus. Some feed on it. That sounds alarming until one remembers that feeding on mucus is not the same as damaging the host. A well-managed mucus layer is continuously produced and continuously remodeled. In a healthy system, mucus-feeders can participate in that remodeling, shaping the niche and influencing which organisms can remain close to the wall.

Akkermansia muciniphila has become a kind of emblem for this interface life. Discovered relatively recently compared with the great pathogens of nineteenth-century bacteriology, it specializes in using mucin, the main component of mucus, as a resource. In animal studies, its presence has been linked to aspects of metabolic health, and its abundance in humans has been associated with certain dietary patterns and physiological states. It is tempting to treat *Akkermansia* as a probiotic mascot, but its deeper interest lies in what it represents: an organism that lives where the host's rules are strictest, helping to define the boundary conditions for the rest of the community. An architect does not merely occupy space; an architect helps decide what kinds of tenants the building can support.

On the worker side of the ledger, a microbe can matter because it produces molecules that the host uses or responds to. One of the clearest examples is the guild of microbes involved in producing short-chain fatty acids (SCFAs), especially butyrate. Butyrate is a small molecule with an oversized résumé. It can be used as fuel by cells lining the colon. It can influence the behavior of immune cells. It can act as a signal, binding to certain receptors on host cells. It can also influence gene expression through mechanisms that, in simplified terms, affect how tightly DNA is packaged and how easily certain genes are switched on or off. The details can be complicated, but the everyday implication is not: dietary fiber becomes biologically meaningful partly because microbes transform it into compounds that the host interprets as instructions.

The key word here is *transform*. Butyrate does not spring from fiber in a single leap. It is often the end product of a chain of microbial cooperation, a metabolic relay race in which the baton is a molecule. This is where guilds become more than a metaphor. Many gut microbes cannot do the entire job of converting a complex plant fiber into butyrate. Some specialize in the first stage: breaking the fiber into smaller fragments. Others specialize in fermentation of those fragments into intermediate products such as acetate or lactate. Still others specialize in turning those intermediates into butyrate. The community output is produced not by a lone genius but by a network.

A cross-feeding guild: cooperation that turns fiber into butyrate

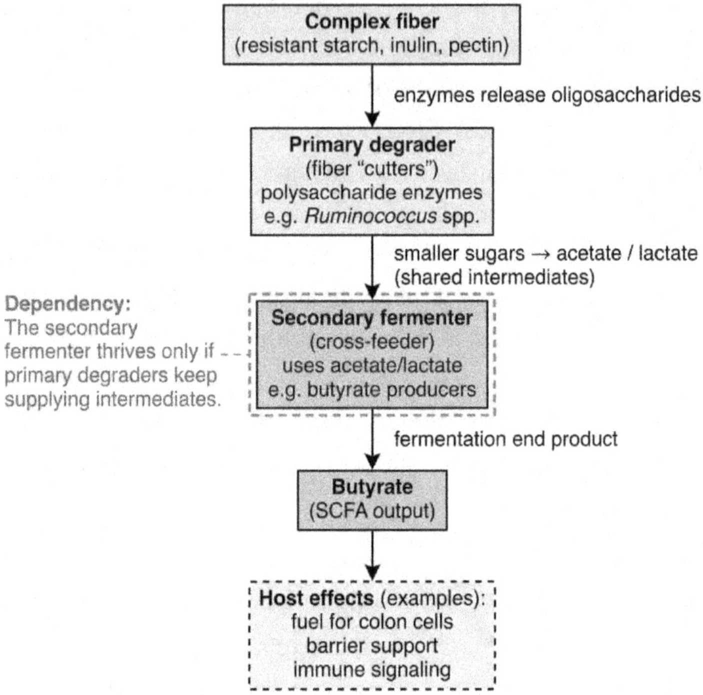

This picture captures something easy to miss when focusing on individual species: the microbiome often behaves less like a set of independent organisms and more like a set of linked industries. If a primary degrader declines, the secondary fermenters that depend on its by-products can also decline, even if they were not directly harmed. The result can be a disproportionate change in community output, a kind of metabolic cascade. When people speak of a microbiome being "knocked out of balance," this is one plausible meaning: not simply that some microbes disappeared, but that a chain of cooperation was interrupted.

A historical anecdote from microbiology makes the point in a different register. For decades, much of gut microbiology was constrained by the practical difficulty of working with strict anaerobes—organisms that cannot tolerate oxygen. Robert Hungate, working in the mid-twentieth century, developed techniques for culturing anaerobic microbes by keeping them away from oxygen during handling, an advance that made it possible to study complex fermenting communities more directly. The technical barrier mattered because oxygen sensitivity is not a quirk; it is a clue. Many of the microbes that do the heavy fermentation work in the colon are anaerobes, living in the low-oxygen depths of the gut where they can specialize in turning complex carbohydrates into SCFAs. If one only studies the oxygen-tolerant organisms, one gets a distorted view of the ecosystem, with too many surface dwellers and not enough workers in the metabolic engine room. Better tools did not merely add more names to the list; they revealed a division of labor that had been there all along.

Within that engine room, some workers have become famous for their apparent relationship to inflammation. *Faecalibacterium prausnitzii* is one such organism: often abundant in healthy guts, sometimes reduced in inflammatory bowel disease cohorts, and studied for its potential immunomodulatory effects. It is easy, and risky, to turn such findings into a simplistic story of "good" bacteria missing from "sick" people. The more careful interpretation is ecological. *F. prausnitzii* belongs to a functional neighborhood: organisms associated with fermentation and butyrate production, living in conditions that tend to coincide with a well-maintained gut barrier and an immune system that is not on constant alert. Inflammation changes the habitat—chemically, physically, immunologically—and habitats select for different residents. The absence of a worker may be both a symptom and a contributor: inflammation makes the job harder, and the missing output may make inflammation easier to sustain.

The architect-worker framing also clarifies why certain therapies are unpredictable. Consider the common intuition behind probiotics: introduce a helpful organism and benefit follows. Sometimes that works, especially when the introduced microbe can temporarily occupy a niche and provide a function the host needs. Often it disappoints, and ecology offers a blunt reason. Introducing a worker into a city without fixing its supply chain does not guarantee productivity. If a butyrate-producing organism arrives but the diet lacks fermentable substrates, or if the community lacks the upstream degraders that provide key intermediates, the worker cannot easily do its job. In the opposite direction, changing the supply chain—increasing diverse fibers, for instance—can sometimes increase the output without importing any new workers, because existing guilds expand when resources become available.

Keystone species make the problem more interesting. If a keystone is missing, the ecosystem may not offer the right niches for other organisms to persist, even if they are introduced. An interface specialist like *Akkermansia* may influence mucus dynamics and the local chemical gradients near the gut wall; other organisms respond to those gradients. A bile-acid-transforming organism may alter the pool of bile acids in ways that affect not only fat digestion but also microbial competition and pathogen resistance. In such cases, a targeted intervention can have broader consequences, because it changes the rules of the habitat rather than merely adding another competitor.

None of this requires imagining the microbiome as a perfectly harmonious cooperative. Cross-feeding is not kindness; it is opportunism with a stable outcome. One organism releases a metabolite because it cannot use it efficiently, or because releasing it helps it keep fermentation moving; another organism consumes it because it can. Cooperation emerges because the chemistry and the environment make it advantageous, not because microbes have agreed on a plan. The host, meanwhile, is not a passive landlord. It controls pH, secretes mucus, releases antimicrobial peptides, and mounts immune responses. These host controls shape which guilds flourish and which wither.

The payoff for thinking in jobs rather than names is a more realistic sense of what "a healthy microbiome" might mean. It is not a museum exhibit with fixed species labels. It is a working ecosystem that can keep doing its essential tasks under the ordinary perturbations of life: missed meals, dietary variation, mild infections, stress, and the occasional medication. Diversity supports that resilience, but guild structure explains it. The key is not merely how many species are present, but whether the community contains the right mix of architects to maintain habitats and the right mix of workers to keep metabolic outputs within a range the host has learned to expect.

A person can go years without thinking about any of this, which is part of the microbiome's strangeness as an "organ." When a heart skips a beat, you feel it. When the pancreas fails, you notice quickly. The microbiome can change profoundly without any immediate sensation at all, and yet the consequences may appear later in the form of altered immunity, vulnerability to infection, or a gut that no longer tolerates the same foods. Much of microbial work is quiet. It happens in the dark, in gradients of oxygen and mucus, as enzymes shave sugars from fibers and as metabolites pass from one organism to another and then into the host.

A useful way to picture the healthy state is not as a perfectly balanced scale but as a well-run marketplace. Many vendors sell overlapping goods. Some specialize, some generalize. The presence of multiple sellers keeps supply reliable when one stall closes. The architects keep the stalls standing and the streets passable. The workers transform raw materials into things the city needs. When that system is intact, the gut does not need to be identical from person to person to be recognizably healthy; it needs to remain capable, day after day, of turning the world's messy inputs into stable, livable outputs.

Chapter 2

Border Control

The human gut is often imagined as a simple pipe, but it is actually a highly guarded frontier where the immune system constantly negotiates with the outside world. We will navigate the delicate architecture of the mucus barrier, separate the marketing hype of "leaky gut" from the medical reality of intestinal permeability, and discover how our cells constantly sample bacterial chatter to distinguish between peaceful residents and dangerous invaders.

2.1 The Demilitarized Zone

A useful way to picture the colon is not as an empty pipe but as a crowded riverbed. The center is a flowing stream of food remnants, shed cells, and microbial life. Along the banks sits a human tissue only one cell thick, richly supplied with blood and nerves, and tasked with absorbing water and salts while refusing entry to the wrong kind of visitor. If the wall of that river were bare, the arrangement would be absurdly fragile: trillions of bacteria on one side, a warm nutrient-rich body on the other, and only a single layer of cells in between.

What makes the arrangement workable is that the body rarely allows microbes to stand face-to-face with living tissue. Instead, it maintains a buffer zone: a carefully engineered separation that is neither a simple fence nor an all-out battlefield. In the colon, that buffer takes a striking physical form. Rather than one undifferentiated slime coat, the mucus is organized into two layers with very different jobs. Malin Johansson and colleagues (2008) helped crystallize this picture by showing that, under healthy conditions, bacteria largely occupy an outer mucus region, while the inner mucus layer adjacent to the epithelium remains effectively bacteria-free. Gunnar Hansson's group later synthesized and expanded this view, emphasizing that mucus is a dynamic, renewable system rather than a static coating.

The point is not to sterilize the gut. Total sterility would be both impossible and undesirable. The point is to enforce distance. Most gut bacteria are not plotting an invasion; they are simply doing what living things do: eating, multiplying, and clinging to surfaces. The body's task is to let them do that in the right place, at the right density, with the right level of intimacy. The mucus layers create a kind of demilitarized zone: a space where microbes can exist, but where direct contact with the epithelium is restricted and carefully negotiated.

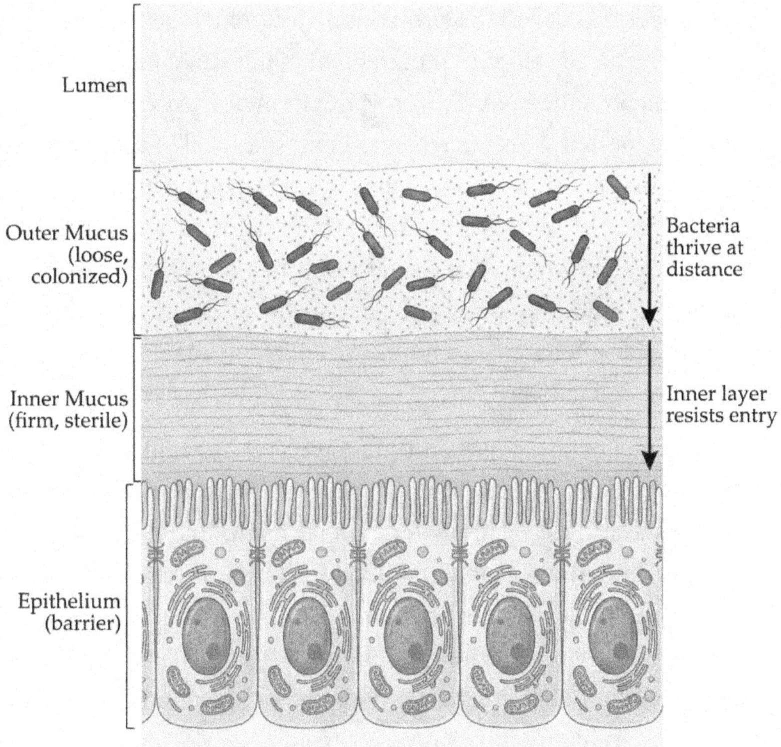

Schematic: Colonic Barrier Cross-Section

The two-layer arrangement matters because bacteria are not merely floating in the gut contents like pepper in soup. Many are adapted to life on surfaces, where food is steady and neighbors are close enough for cooperation and competition. The outer mucus layer provides exactly that: a semi-structured habitat, rich in complex sugars, constantly replenished, and thick enough to host dense communities. If you imagine a coral reef, you will be tempted to picture something rigid; mucus is not rigid at all. Yet it can be surprisingly organized. In the colon, the outer layer is looser and more easily stirred by the motion of the gut. It is a place where bacteria can live and feed, including on the mucus itself.

That last fact can sound unsettling until you remember what mucus is made of. The key structural ingredient in intestinal mucus is a gel-forming mucin called MUC2 (or *Muc2* in mice). Mucins are large proteins decorated with many sugar chains, a bit like a bottlebrush where the bristles are carbohydrates. Those sugar chains do several things at once. They attract water, which helps mucus form a hydrated gel. They create a physical mesh that slows the movement of particles. And they offer a diverse menu of chemical structures that particular microbes can nibble. In other words, mucus is both architecture and cuisine.

The inner mucus layer plays a different role. In healthy colon it is denser, more firmly attached, and arranged so that bacteria have difficulty penetrating it. This is the crucial point: the body does not rely on the epithelium as the first line of defense. It relies on a *space* maintained in front of the epithelium. The inner mucus layer is, in effect, a no-parking zone. The epithelium can perform absorption and sensing without being constantly bumped, prodded, and chemically provoked by living microbes pressed against it.

How does mucus acquire such different personalities in two adjacent layers? Part of the answer lies in physics and chemistry: how MUC2 polymers expand, cross-link, and are modified after secretion. Another part lies in time. Mucus is not poured once and left to harden. It is continuously produced and continuously eroded. The outer layer is, in a sense, the "used" layer: it has been exposed to digestive enzymes, microbial enzymes, and mechanical shear. It is also where the microbial community is densest, and microbes are extraordinarily good chemists. Given time, they can loosen, clip, and remodel the environment around them. The inner layer, by contrast, is newer, less degraded, and maintained in a state that resists colonization.

This constant renewal is orchestrated by goblet cells, specialized epithelial cells named for their goblet-like shape. Under the microscope, a goblet cell looks as if it is filled with pale bubbles. Those bubbles are mucin granules, packages of mucus components ready to be released. When conditions are calm, goblet cells secrete mucus steadily, like a slow-release irrigation system. When conditions are threatening, secretion can become dramatic and collective, as if a neighborhood has agreed on a synchronized closing of shutters.

The existence of "alarm" modes is not just metaphor. In mouse studies, a subset of goblet cells located near the openings of colonic crypts has been described as *sentinel* goblet cells. These cells appear capable of detecting microbial patterns through receptors that recognize common bacterial molecules. When triggered, they can initiate a burst of MUC2 secretion, and even propagate a wave of secretion to nearby goblet cells via cell-to-cell communication. The details are still being worked out in humans, and the word "sentinel" should not be taken as a literal job title stamped on a cell. But the larger idea is robust: mucus secretion is regulated, responsive, and tied to innate immune sensing pathways such as inflammasomes (including NLRP6 in several mouse models). A barrier that can sense is a barrier that can negotiate. It can tighten the boundary when contact becomes risky, rather than reacting only after invasion has occurred.

The demilitarized zone is also chemical. Even if you built a perfect physical filter, bacteria and their fragments would still drift toward the epithelium. The body therefore layers antimicrobial molecules into the mucus environment, but does so with restraint. Some antimicrobial peptides act like molecular spears: they can puncture bacterial membranes. Others bind nutrients so tightly that bacteria cannot access them. In the small intestine, Paneth cells at the base of crypts are famous for producing antimicrobial peptides, shaping gradients along the crypt-villus axis. The colon has fewer Paneth cells, but it still uses antimicrobial strategies, and the bilayer mucus arrangement helps keep those strategies where they are most useful.

The goal is not to bathe the entire lumen in antibiotics, which would be self-defeating. The goal is to keep the immediate neighborhood of the epithelium relatively quiet.

A concrete example makes this logic easier to feel. *Clostridioides difficile* infection, often following antibiotic use, is a reminder that the gut is an ecosystem with rules. Antibiotics can reduce microbial diversity and remove competitors that normally keep certain organisms in check. When *C. difficile* expands, it can produce toxins that injure epithelial cells and inflame the gut. But the damage is not only a matter of "bad bacteria present." It is also about geography. When inflammation disrupts mucus production, mucus structure, or the balance of chemical defenses, bacteria that were previously held at a respectful distance can approach the epithelium, increasing the intensity of immune activation. Treatments like fecal microbiota transplantation work, in part, by restoring community-level competition and metabolic functions that help re-stabilize the habitat. In everyday language, they help repopulate the outer zone with a community that makes the border easier to police.

This geography also helps explain why the colon can tolerate an enormous bacterial load without erupting into constant inflammation. The immune system does not need to "ignore" bacteria in some mystical way. It can afford to be calm because, much of the time, bacteria are not physically pressing against the cells that would trigger the most alarm. Many immune receptors are positioned so that they respond strongly to bacteria that cross the line, not to those living peacefully at a distance. The mucus layers, by enforcing that distance, reduce false alarms.

There is a historical footnote here worth savoring because it shows how biology sometimes hides in plain sight. Mucus was long treated as a nuisance: something to be rinsed away so the "real" tissue could be examined. In many lab protocols, mucus was removed or dissolved during sample preparation. When researchers began to preserve it carefully and visualize it in its native state, a different landscape appeared. The discovery that the colon's mucus is layered, with distinct microbial localization, is a reminder that our picture of the body depends on what we are willing to keep in the frame. A barrier that looks thin and unimpressive after harsh processing can, in life, be thick, structured, and full of rules.

Those rules are not the same everywhere. The small intestine faces a different challenge than the colon. It sees more incoming food molecules, including fats that need to be absorbed, and it has a faster flow that tends to wash microbes downstream. Its mucus is typically a single layer and is often less firmly attached. The colon, in contrast, is a fermentation chamber where microbes have time to settle and multiply. It reclaims water, slowing the contents further and making surface living easier. The colon therefore invests more heavily in a stable, layered mucus system. The same design principle appears in different architectures: match the border to the traffic.

Diet quietly participates in this border control. Many gut microbes prefer dietary fiber, the complex carbohydrates our own enzymes cannot digest. When fiber is plentiful, bacteria can feast on incoming plant polysaccharides and on each other's byproducts. When fiber is scarce, some microbes increase their use of mucus glycans as an alternative food source. This does not automatically mean mucus is "eaten away" in a simplistic sense; the system is resilient and replenished. But it does mean that what you feed the microbial community influences how hard it pulls on the body's own infrastructure. The demilitarized zone is a shared space, maintained by the host but inhabited and modified by microbial residents.

It is tempting, in popular discussions, to talk as if the gut has two possible states: either a perfect wall that keeps everything out, or a disastrous breach where toxins and bacteria flood the body. Real barriers are rarely so binary. The mucus layers can be thicker or thinner, more or less penetrable, and renewed at different rates depending on diet, infection, inflammation, and even the rhythms of daily life. Goblet cells can be stressed, depleted, or hyperactive. Microbes can shift from cooperative to competitive modes. The border is always under negotiation.

Yet the central achievement remains astonishingly stable across most healthy lives. A human can carry a microbial biomass in the colon that rivals the weight of the brain, and still keep those microbes a measured distance away from living tissue for decades. That is not accomplished by a single trick. It is accomplished by stacking partial solutions: a physical gel that slows approach, an inner layer that resists penetration, a renewal system that can surge when threatened, and a chemical environment that discourages microbial trespass without sterilizing the neighborhood. When the system works, it feels like nothing at all. That quiet is not the absence of activity; it is the sound of a border that is being maintained every minute.

2.2 The Myth and Reality of ``Leaky Gut"

The phrase "leaky gut" has a certain tabloid genius. It sounds concrete, slightly alarming, and wonderfully explanatory. Fatigue? Brain fog? Skin problems? Joint pain? The gut is leaking. Fix the leak, and the story promises to tie together a messy bundle of symptoms with a single piece of plumbing.

There is a real scientific idea hiding inside that slogan, but it is narrower, more technical, and less universal than the wellness version. Clinicians and physiologists usually talk about *intestinal permeability*: how readily substances pass from the gut lumen into the body. Permeability is not inherently bad. A gut that allowed nothing through would be useless. The question is *what* is passing, *where* it is crossing, and *under what conditions*.

The temptation to make permeability a master explanation is understandable because the gut really is a border. The body is separated from a microbial metropolis by a single-cell-thick epithelium, and the colon's mucus layers keep most microbes at a respectful distance. But even with that demilitarized zone, the border is busy. Water, salts, vitamins, fatty acids, and small fragments of digested food must cross every day in vast quantities. If you imagine the gut lining as a static wall, then any crossing sounds like a breach. If you imagine it as a living border crossing with gates, guards, and paperwork, then crossing becomes normal—and the drama lies in the exceptions.

The most important "gate" is not a hole punched through a cell, but the seam between neighboring cells. Intestinal epithelial cells sit shoulder to shoulder like tiles. The space between them is sealed by structures called *tight junctions*: multiprotein complexes that act as adjustable gaskets. Under the microscope they are not visible as a single latch you could point to; they are a belt-like region where proteins such as claudins and occludin, connected to scaffolding proteins like ZO-1, link one cell to the next and to the cell's internal skeleton. Tight junctions can be remodeled. Some claudins form pores that selectively let small ions pass; others tighten the seal. The junction is less like mortar and more like a zipper with adjustable teeth.

This is where the wellness narrative often goes wrong. It treats the gut as if it has two states: sealed (healthy) and torn open (diseased), with toxins pouring through. Real physiology is a spectrum and, crucially, a *regulated* spectrum. The gut changes permeability in response to food, microbes, hormones, stress signals, and immune mediators. During an infection, loosening the barrier slightly can help flush water into the lumen and speed the removal of pathogens, even though it may temporarily allow more microbial molecules to approach the immune system. In other contexts, tightening the barrier can reduce immune stimulation. The border is not a wall that occasionally breaks; it is a living interface that can be tuned, sometimes wisely and sometimes at a cost.

If tight junctions are adjustable, what turns the dial? One well-studied pathway involves a protein called *myosin light chain kinase* (MLCK). MLCK influences the contractile machinery inside cells—the same kind of actin-myosin system that lets muscles contract, but here used on a microscopic scale. When MLCK activity rises, the perijunctional actomyosin ring can tighten like a drawstring around the top of the cell, pulling on junctional complexes and increasing paracellular permeability (movement between cells). Inflammatory cytokines—chemical signals released by immune cells and stressed tissues—often push in this direction. Tumor necrosis factor (TNF) and interferon-gamma, for example, can induce junctional remodeling in ways that make the epithelium more permissive. Another mechanism involves the internalization (endocytosis) of junctional proteins such as occludin, changing the local architecture of the seal. These are not exotic details; they are reminders that a shift in permeability is frequently an *active cellular response*, not simply damage.

So what does "increased intestinal permeability" mean in practice? It depends on the context and the measurement. Researchers have used a variety of assays, but a classic approach in humans is the dual-sugar test, often described as the lactulose–mannitol test. The person drinks a solution containing two sugars. Mannitol is small and can pass more readily through the intestinal barrier; lactulose is larger and typically crosses less. By measuring how much of each appears in urine over a set time, clinicians and researchers infer something about barrier function. A higher lactulose-to-mannitol ratio can suggest increased permeability. The method has value, but it also has limitations: results can be influenced by kidney function, gastric emptying, intestinal transit time, and the fact that different gut regions have different permeability properties. It is also not a single universally standardized clinical test deployed in every hospital for every complaint. It is a tool used carefully in particular research and clinical settings.

If that sounds less satisfying than a simple blood test labeled "leaky gut: yes/no," that is because biology rarely cooperates with marketing. The wellness industry has tried to fill this gap with biomarker tests, and one name appears again and again: *zonulin*. Zonulin is often described as a master regulator of tight junctions, and it is true that work on zonulin helped popularize the idea that the barrier could be physiologically modulated, not merely broken. The trouble is that the way zonulin is measured in many commercial and even research contexts has been shaky. Scheffler and colleagues (2018) examined a widely used commercial "zonulin" ELISA (an antibody-based assay) and found evidence that it did not detect the expected precursor of haptoglobin-2 (the molecule many refer to as zonulin), and instead likely recognized other proteins such as properdin. If you build an entire story about permeability on a test that may be measuring something else, the story can acquire an undeserved aura of precision. Numbers printed on a lab report feel authoritative even when the underlying measurement is ambiguous.

A historical analogy helps. In the early days of endocrinology, crude assays sometimes detected "something" in blood and gave it a hormone's name because it correlated with a condition. Later, improved methods revealed that the assay had been picking up a related molecule or even a mixture. Science corrected course, but the names and narratives sometimes lingered in public imagination. Zonulin has become, for some consumers, what "adrenal fatigue" is for others: a label that feels like an explanation, attached to tests and supplements that may not map cleanly onto physiology.

None of this means permeability is imaginary. It means it is specific. There are diseases where barrier disruption is central rather than incidental. Celiac disease provides a concrete example because it has a clear trigger (gluten, specifically components like gliadin), a clear immune mechanism, and tissue changes that can be seen directly in biopsies. In celiac disease, ingestion of gluten in susceptible individuals triggers an immune response that damages the small intestinal lining, leading to villous atrophy (flattening of the absorptive surface) and malabsorption. Barrier changes occur in this inflammatory context, and permeability can be altered. The key point is not that "gluten makes everyone's gut leak." The key point is that in a defined disease state with a defined immune reaction, barrier regulation becomes clinically relevant.

Inflammatory bowel diseases, such as Crohn's disease and ulcerative colitis, also sit close to the core of this topic, though their causes are more complex and involve genetics, immune regulation, microbiome composition, and environmental triggers. In these diseases, the border becomes a battleground. Tight junctions are affected, mucus can be altered, and ulcers can form—true breaches rather than mere loosened seals. Here the permeability question is not a wellness metaphor; it is entangled with bleeding, pain, diarrhea, and long-term risk.

Between the dramatic world of inflammatory disease and the sweeping claims of wellness culture lies a subtler phenomenon that has attracted serious research attention: *metabolic endotoxemia*. The term sounds forbidding, but the idea is straightforward. Many gut bacteria, especially Gram-negative species, have lipopolysaccharide (LPS) in their outer membrane. LPS is sometimes called an endotoxin because, when it reaches the bloodstream in sufficient amounts, it can trigger inflammatory signaling through receptors like TLR4. The body is not indifferent to LPS; it interprets it as a microbial fingerprint.

In metabolic endotoxemia, small increases in circulating LPS (or LPS activity) are proposed to contribute to low-grade systemic inflammation associated with obesity, insulin resistance, and related metabolic disorders. This is not the same as sepsis, where massive amounts of bacterial products overwhelm the body. It is more like a persistent smoky haze rather than a blazing fire. The permeability link comes from the plausible route by which LPS could cross from the gut lumen into the body more readily under certain conditions—for example, when dietary patterns, bile acids, microbial community structure, and barrier regulation all shift in ways that allow more microbial fragments to reach immune cells.

The crucial word is *plausible*. In humans, establishing causality is difficult. Higher markers of inflammation can correlate with measures that suggest altered barrier function, but correlation can run in several directions. Inflammation itself can loosen tight junctions. Diet can reshape microbes and simultaneously affect host physiology. Genetic differences can influence immune tone and barrier proteins. The temptation is to point to a single villain—"your gut is leaking toxins"—but the more honest picture is an interacting system with feedback loops. Tight junctions are not only victims; they are participants in an immune-metabolic conversation.

A concrete, everyday example shows how easy it is to overinterpret a real mechanism. After a night of heavy drinking, some people feel physically inflamed: flushing, poor sleep, a fast heartbeat, gastrointestinal upset. Alcohol can affect the gut barrier and the microbiome, and in animal models it can increase permeability. It is tempting to claim that the hangover is simply "toxins leaking from your gut." But alcohol also directly affects the brain, the liver, sleep architecture, hydration status, and stress hormones. The gut may be part of the story without being the whole plot. The body is not a single pipeline;

it is a network.

Why does the myth persist so powerfully? Because it offers moral clarity. The wellness version of leaky gut usually comes packaged with a villain list (processed foods, lectins, dairy, gluten, sugar, seed oils, stress, modern life) and a hero list (supplements, cleanses, bone broth, collagen, proprietary powders). The story is psychologically satisfying: it turns diffuse suffering into a fixable boundary problem. It also leverages a real intuition: borders matter. People can feel, correctly, that their gut is connected to their skin, mood, and energy. The mistake is not sensing connection; it is believing that a single diagnosis explains every connection.

A more grounded way to think about intestinal permeability starts by recognizing that the gut lining has multiple layers of border control. The mucus creates distance. The epithelium forms a cellular wall. Tight junctions tune the seams. Immune cells patrol beneath the surface. Blood vessels and lymphatics decide what happens to absorbed molecules. When permeability changes, it may do so at any of these levels, and the consequences depend on what else is happening: the microbial community, diet composition, and the immune system's current temperament.

This perspective also changes what "fixing the barrier" should mean. If someone has a specific disease—celiac disease, inflammatory bowel disease, a serious infection—the right response is diagnosis and evidence-based treatment, not a generic supplement stack. If someone is generally unwell, with nonspecific symptoms, the right response is often slower and less theatrical: rule out major conditions; consider sleep, diet quality, medications, alcohol, and stress; and be cautious about tests that promise a single number as a master key. The gut is not a brittle wall that fails catastrophically in millions of people without leaving clear medical footprints. It is a responsive, metabolically active interface that can be strained, inflamed, and remodeled, sometimes subtly, sometimes dramatically.

A final irony is worth holding onto because it keeps the topic honest. The body sometimes loosens the border *on purpose*. During an immune response, a slightly more permissive barrier can increase sampling of luminal contents, helping the immune system learn what it faces. During nutrient absorption, the barrier must allow passage of an enormous daily flow of molecules. The goal is not maximal tightness; the goal is appropriate selectivity. Too much permeability in the wrong context can drive inflammation. Too little permeability, or an overzealous immune response to harmless molecules, can also cause disease. Health is not achieved by sealing the gut shut; it is achieved by keeping the gates well-run, the seals well-maintained, and the alarms calibrated so they ring for genuine threats rather than for the ordinary bustle of life inside the lumen.

2.3 Diplomatic Signals

A paradox sits at the heart of the gut. The immune system evolved to recognize and destroy microbes, yet the intestine is home to a microbial population that outnumbers human cells and carries a collective gene set far larger than our own. If the immune system treated every bacterial encounter as an emergency, the result would be constant inflammation, a kind of immunological civil war fought on the intestinal surface. If it treated microbes with indifference, true pathogens would have an open door. Peace in the gut is not achieved by ignorance; it is achieved by discrimination—the ability to respond differently to different kinds of microbial contact, and to do so without causing collateral damage to the tissue that has to absorb food tomorrow.

That discrimination begins with a simple fact of geography. Most gut microbes do not sit directly on epithelial cells. They live in the lumen and, in the colon, in the looser outer mucus layer. The inner mucus and the epithelial surface are closer to a protected embassy district than an open marketplace. Yet the immune system still needs information. It cannot base policy on rumors. It has to sample what is in the gut, interpret it, and decide whether the current residents are peaceful neighbors, opportunistic looters, or invaders arriving with weapons.

The intestine therefore runs an unusually sophisticated surveillance program. It does not rely on random leaks. It uses designated ports of entry for information, with rules about what can pass and in what form. The elegance of these arrangements can be missed if one imagines the immune system as a set of soldiers patrolling for enemies. Much of gut immunity is closer to customs and diplomacy: checking papers, escorting visitors, and keeping communication channels open so that misunderstandings do not turn into violence.

A central character in this story is the dendritic cell. Dendritic cells are professional antigen-presenting cells: they pick up fragments of microbes or food proteins (antigens) and show them to T cells, the immune system's decision-makers. In many tissues, dendritic cells are like detectives arriving after a break-in. In the gut, they are more like border officials, stationed where they can continuously collect samples.

How do they sample a world separated from them by epithelium and mucus? One answer involves physical reach. Certain dendritic cells can extend thin projections between epithelial cells to access material near the lumen while keeping the epithelial barrier largely intact. The image can be surprising the first time you see it in a diagram: a cell from the immune side threading a slender "arm" upward, as if peering through blinds. This reaching behavior makes sense if you think of the epithelium not as a fortress wall with no communication, but as a living partition whose seams can accommodate controlled interactions.

Another, even more controlled route uses the epithelium itself as a sampling device. In the small intestine, goblet cells—better known for producing mucus—can form what have been called goblet cell-associated antigen passages (GAPs). Jeremiah McDole and colleagues (2012) described how goblet cells can deliver soluble antigens from the lumen to dendritic cells in the tissue below, particularly to a subset often associated with tolerance ($CD103^+$ dendritic cells). The word "tolerance" here does not mean passivity. It means an active program of restraint: teaching the immune system that certain recurring antigens are not threats worth attacking.

This arrangement is emotionally counterintuitive. The immune system is often imagined as a paranoid security system that should never allow outsiders in. Yet the gut, which has perhaps the most complicated outside-inside boundary in the body, deliberately transports luminal material across that boundary. It does so because reliable peace requires reliable information. A border that never processes travelers would not be safe; it would be blind. The problem is not entry *per se*; it is entry without regulation.

The most famous organized port of entry is the Peyer's patch, a specialized lymphoid structure embedded in the small intestine. Peyer's patches are like consulates: places where immune cells gather, exchange information, and initiate responses. Above each Peyer's patch is a region of epithelium enriched in M cells (microfold cells). M cells are specialized for transcytosis, the transport of particles and antigens from the lumen to immune cells beneath. Their surface is less geared toward absorption and more toward sampling. Pathogens have learned to exploit M cells, which is one reason they are studied so closely, but in everyday life M cells are also a normal route by which commensal-derived antigens are introduced to the immune system in a controlled fashion.

A striking detail, described in carefully designed mouse studies, is that dendritic cells can sample through M-cell-associated structures in a coordinated way, extending dendrites through pores linked to M cells. That is a mouthful, but the logic is plain: the epithelium and immune system collaborate. The border is not simply defended from below; it is co-managed.

Why go to such trouble? Because the output of this sampling is not always attack. One of the gut's signature immune products is secretory IgA (SIgA), an antibody designed for mucosal surfaces. Antibodies are often framed as bullets: lock onto a microbe, recruit immune cells, trigger destruction. IgA is different in personality. In many contexts it behaves less like a bullet and more like a set of handcuffs or a layer of insulation. It can bind bacteria and prevent them from attaching too closely to the epithelium. It can neutralize toxins. It can clump microbes together (agglutination), changing their ability to move and colonize. Importantly, IgA can do these things without triggering the same degree of inflammatory firepower associated with other antibody classes.

This matters because inflammation is expensive. It damages tissue, changes permeability, and disrupts the very border it is meant to defend. In the gut, chronic high-grade inflammation can flatten villi, erode mucus systems, and create a feedback loop where barrier disruption encourages further immune activation. IgA helps avoid that trap by shaping microbial behavior in a comparatively nondestructive way. It is one of the immune system's tools for maintaining a crowded neighborhood without constant raids.

There is a concrete, human-scale reason to care about IgA that has nothing to do with abstract immunology: infants. A newborn's immune system has to learn, quickly, how to live with microbes. Breast milk contains antibodies, including IgA, that help coat microbes and guide early microbial colonization. In effect, maternal immunity provides a temporary diplomatic corps while the infant's own immune institutions are still under construction. When people speak about the microbiome's importance in early life, this is part of what they mean: it is not only about which microbes arrive, but about the immune context in which they are introduced.

The idea that antibodies could protect without provoking inflammation was not always obvious. Early immunology was shaped by dramatic diseases: diphtheria, tetanus, pneumonia, bloodstream infections. Those battles happen in tissues where inflammation can be lifesaving despite collateral damage. Mucosal surfaces forced immunologists to confront a different world. The gut is constantly exposed to antigens from food and microbes, and the body cannot afford to respond to all of them with the full violence of systemic immunity. The discovery and characterization of IgA as a dominant mucosal antibody class, and the realization that it is actively transported across epithelial cells into the lumen as secretory IgA, helped reveal a distinct style of immune protection. The gut does not simply use less immunity; it uses *different* immunity.

This also helps clarify a frequent misconception: tolerance is not weakness. If the immune system attacked commensals relentlessly, it would destabilize the microbial ecosystem and injure the host tissue, making true infections more likely. Restraint is not a lapse in defense; it is a strategy for long-term defense in a high-exposure environment.

The immune system's diplomatic style in the gut is reinforced not only by antibodies and sampling routes but by chemical messages exchanged with microbes. Among the most important of these are short-chain fatty acids (SCFAs), especially acetate, propionate, and butyrate, produced when microbes ferment dietary fiber. These molecules are not mere waste products. Butyrate, in particular, is a major fuel for colon epithelial cells and acts as a signaling molecule that can influence gene expression and immune programming. In several lines of research, SCFAs have been shown to bias immune responses toward regulation, including promoting the development or function of regulatory T cells (Tregs), which help prevent excessive immune activation.

This is one of the ways the microbiome behaves like an organ: not simply by occupying space, but by producing metabolites that circulate and alter host physiology. A gut ecosystem that produces abundant butyrate is not just digesting leftovers; it is shaping the tone of the immune system. If the microbial community shifts away from fiber fermentation—because of diet, antibiotics, or disease—the immune consequences can follow.

A vivid example of how diplomacy can fail involves the aftermath of antibiotics. Antibiotics can be lifesaving, but they are also blunt instruments. They can reduce microbial diversity and disrupt the microbial networks that normally keep opportunists in check. When the community destabilizes, the immune system receives different signals: fewer SCFAs, altered bile acid metabolism, different microbial fragments. At the same time, niches open up for organisms like *Clostridioides difficile*, which can cause severe colitis. The immune response in *C. difficile* infection is not only about killing the pathogen; it is also about coping with toxins and tissue injury. Inflammation rises, the epithelium suffers, and the carefully maintained separation between microbes and tissue becomes harder to enforce. What looked like a simple infection becomes a systems problem at the border, where ecology and immunity spiral together.

The diplomatic framing also sheds light on why the gut immune system is sometimes implicated in diseases far from the intestine. If microbial products and metabolites influence immune tone, then changes in the gut ecosystem can plausibly shift the baseline activation state of immune cells elsewhere. This does not mean the microbiome is a single master switch for every condition, but it does mean that the gut can act like a central negotiating table: a place where the body continually decides how trigger-happy or restrained it should be.

At the level of everyday experience, this helps explain why the gut is so emotionally salient. People often report that stress affects digestion, and that digestive disturbances affect mood. Part of that is neural and hormonal. Part of it is immune. An inflamed gut feels different because the immune system's chemical language is not confined to the intestine; cytokines and metabolites can influence the brain and the rest of the body. When the gut immune system maintains peace, it is doing more than preventing diarrhea. It is helping set the body's inflammatory thermostat.

The elegance of mucosal diplomacy is that it can be specific without being destructive. A pathogen is not defined only by its species name; it is defined by behavior and context. Some bacteria are harmless residents in one setting and dangerous invaders in another. The immune system's job is to integrate signals: how close microbes are to the epithelium, whether they are breaching mucus, whether epithelial cells are stressed, what metabolites are present, what pattern-recognition receptors are being engaged, and what kind of tissue damage is occurring. Responses can range from quiet IgA coating and tolerogenic education to the recruitment of inflammatory cells and the production of antimicrobial factors.

The gut border is therefore neither naive nor perpetually at war. It is a negotiated boundary, maintained by physical separation, regulated sampling, and immune tools designed to control without scorched earth. When the system works well, most of the day's microbial drama ends not with explosions but with paperwork: antigens sampled, antibodies secreted, microbes nudged into safer zones, epithelial cells fed by microbial metabolites, and immune cells instructed to keep their weapons lowered unless a real threat appears.

Chapter 3

First Inheritance

We typically view our biological inheritance as a fixed set of genes, yet a second, living legacy is bequeathed to us the moment we enter the world. This narrative traces the microbiome's lifecycle from the chaotic 'land grab' of pioneer species during birth and the chemically coded instructions of breast milk, to the ecological unraveling that characterizes frailty in later years.

3.1 Seeding the Landscape

A human birth is a dramatic change of address. For months, a fetus develops in an environment buffered from the ordinary microbial traffic of the outside world. Then, within minutes, an infant moves from warm, fluid, and relatively sheltered conditions into air, hands, cloth, metal instruments, and skin. That transition is not just psychological and physiological; it is ecological. A new habitat appears: a warm, nutrient-rich gut with an immune system learning the difference between friend and foe, and with surfaces (mouth, skin, airway) that will soon host dense microbial communities. The first organisms to arrive do not merely take up space. They alter the environment in ways that shape who can live there later.

Ecologists have a phrase for this: a founder effect. In island biology, the first birds or insects that reach a newly formed island can set the tone for decades. They claim the best niches, consume key resources, and, sometimes, transform the landscape itself. The infant gut is not an island rising from the sea, but it behaves like one in an important sense: early arrivals have disproportionate influence because the habitat is still being built. Gut oxygen levels, pH, immune tolerance, and even the chemical menu available to microbes are all in flux. Timing matters. Order matters.

The earliest colonizers tend to be organisms that can tolerate oxygen. That may sound odd for a place famous for being anaerobic (low in oxygen), but a newborn gut is not the adult gut. Oxygen diffuses from the gut wall and arrives with swallowed air, and the first microbes include those that can live with it: facultative anaerobes such as *Escherichia coli*, *Enterococcus*, and *Staphylococcus*. Their metabolism consumes oxygen and begins to lower the redox potential, gradually making conditions friendlier for strict anaerobes. This is microbial succession: one wave preparing the ground for the next.

Now comes the crucial twist. The first wave is not drawn from a universal microbial pool. It is drawn from wherever the infant passes and whatever touches the infant during those first minutes and hours. That is why delivery mode has such a powerful signature. Vaginal birth and Caesarean section (C-section) are both forms of delivery, but they are ecologically distinct events.

During vaginal delivery, an infant is pressed against mucosal surfaces and fluids rich in microbes adapted to that environment. A healthy vagina is often dominated by *Lactobacillus* species, organisms that thrive on glycogen-derived sugars and produce lactic acid, helping maintain a low pH that discourages many pathogens. The infant's skin and mouth are exposed to those communities, and the gut receives them indirectly through swallowing and through the cascade of contact that follows. In the simplest picture, the newborn is "painted" with a maternal microbial signature.

During a C-section, the route is different. The infant is delivered through an abdominal incision in an operating theatre, where skin microbes, gloves, drapes, and a carefully managed antiseptic environment dominate what the infant encounters. The newborn still receives microbes (life without microbes is not an option), but the initial set is more likely to resemble skin-associated organisms such as *Staphylococcus* and *Corynebacterium*, along with hospital-associated strains. Many C-sections also involve prophylactic antibiotics given to the mother, and those drugs can reduce the number of microbes available to transfer and can suppress susceptible early colonizers in the infant.

The idea that delivery mode leaves a microbial signature became difficult to ignore once researchers began sequencing microbial DNA directly from infant samples. Around 2010, work led by Maria Dominguez-Bello and colleagues compared microbial communities across newborn body sites shortly after birth and reported a striking pattern: vaginally delivered infants tended to resemble the maternal vaginal microbiota, while C-section infants more closely resembled skin-associated communities. The point was not that C-section infants lacked microbes (they do not), but that the *identity* of the pioneers differed.

The founder effect is compelling because it offers a mechanism rather than a mere association. Suppose a newborn gut is initially seeded with organisms that can rapidly consume oxygen and produce acids that favor a later wave of beneficial anaerobes. Suppose, instead, the pioneers are organisms less suited to that task, or organisms more typical of dry skin than of mucosal surfaces. The difference could shift the speed and direction of succession. In ecological terms, you can think of early colonizers as engineers: they change the habitat, and habitat change changes the rules of competition.

What makes this more than a tidy story is that infant gut communities do not simply drift randomly toward an adult state. They follow a developmental trajectory. Large longitudinal studies using metagenomics have shown that the first year of life features recognizable stages, with abrupt transitions when diet changes and when breastfeeding ends. Delivery mode does not freeze the microbiome into two permanent types, but it can nudge the starting conditions. In many infants, those nudges are temporary; in others, they can be detected months later, especially when combined with other pressures such as antibiotics, prematurity, or prolonged hospital stays.

A helpful way to hold these ideas without oversimplifying them is to separate exposure from retention. Exposure is what the infant meets. Retention is what successfully colonizes and stays long enough to influence the next wave. Newer work has suggested that some key differences in C-section infants may be less about a complete lack of exposure and more about instability: certain microbes may be encountered but fail to establish a durable foothold. *Bacteroides*, for example, often appears later or less consistently in C-section infants in some cohorts, and this may reflect the difficulty of establishing stable lineages without the usual maternal-to-infant transfer routes or under antibiotic pressure. In other words, the early gut can be seeded, but the seeds may not take.

That distinction matters because it changes how one thinks about intervention. If the issue were simply "missing bacteria," then one might imagine that a brief exposure to maternal microbes could restore the typical pattern. If the issue involves retention and ecological context, then a one-time exposure may not be enough, and in some cases could even be risky.

This is where a concrete episode from the history of medicine is clarifying, because it shows how childbirth, microbes, and clinical practice have long been intertwined, even before anyone could see a bacterium.

In the mid-nineteenth century, Ignaz Semmelweis worked at the Vienna General Hospital, where two maternity wards had sharply different mortality rates from puerperal fever, a devastating postpartum infection. One ward was staffed by physicians and medical students, the other by midwives. Semmelweis noticed that the physicians often came directly from autopsies to the delivery room. He hypothesized that some transferable material on their hands was responsible. When he instituted handwashing with chlorinated lime, mortality plummeted. Semmelweis did not know about bacteria in the modern sense; germ theory was not yet widely accepted. Yet his intervention was, in effect, a dramatic reshaping of the microbial exposures surrounding birth.

Semmelweis is often remembered as a tragic hero of infection control, and rightly so. But his story also serves as a reminder that the microbial context of delivery is not a philosophical curiosity. It is a powerful determinant of health. The microbial "seeding" of a newborn is not just about beneficial colonizers; it is also about avoiding harmful ones. The same mechanisms that allow a mother to pass along friendly lineages could, under different circumstances, pass along pathogens. This is why clinical caution matters whenever someone proposes to "restore" a vaginal-like microbial exposure after C-section.

The desire to mimic vaginal exposure has led to the controversial idea of vaginal seeding: transferring maternal vaginal fluids to the newborn after a C-section, typically by swabbing the infant's mouth, face, or skin. The concept is straightforward, almost elegant, and that simplicity is part of its appeal. But biology rarely rewards elegance alone. Major obstetric organizations have urged that vaginal seeding should not be performed outside research protocols because of the risk of transmitting pathogens such as Group B streptococcus, herpes simplex virus, *Chlamydia trachomatis*, or *Neisseria gonorrhoeae*.

The ecological motivation is understandable; the clinical risk is real.

One can also see why the "paint the baby with vaginal microbes" picture is incomplete. The microbes that matter most for long-term gut colonization may not be primarily vaginal. Maternal gut and rectal strains may be more important sources for some key intestinal colonizers, and those strains might be transferred through contact patterns around birth and early life, not solely through passage through the birth canal. If that is true, then the critical feature of delivery mode is not only which microbes touch the infant, but how hospital practices, antibiotics, and early feeding shape the ability of those lineages to establish.

Even without intervention, the microbiome of a newborn changes rapidly. Within days, the gut becomes a battleground of growth rates. Milk arrives, providing lactose, fats, proteins, and a host of molecules that the infant may not digest fully but that microbes can use. The immune system begins sampling microbial antigens and learning which patterns are normal. The gut lining thickens and produces mucus. This is not a passive container being filled; it is a developing organ interface negotiating its first relationships.

The striking feature of this early period is how much of it occurs before an infant has any say in the matter. Adults choose diets, travel destinations, hygienic practices, and medications, and these choices shape the microbiome. Newborns inherit an initial microbial world through circumstances: the mode of delivery, the hospital environment, antibiotic exposures around labor, and the first feeding decisions. "Inheritance" here is not primarily genetic; it is ecological and relational.

The founder effect does not mean that vaginal birth is always microbiologically "better," nor that C-section inevitably produces harm. C-sections save lives. They can prevent catastrophic outcomes in obstructed labor, placenta previa, fetal distress, uterine rupture, and many other emergencies. A C-section is not an ecological error; it is often a medical necessity. The point is subtler: because the early gut is unusually malleable, delivery mode is one of the strongest predictors of which microbes arrive first and which manage to settle.

It is also why simplistic claims can mislead. Some popular accounts imply that the womb is sterile and the newborn is a blank slate. Others, reacting against that idea, claim that microbes colonize the fetus before birth and that birth is less important. The reality is messier. Detecting microbial DNA in low-biomass tissues is technically difficult, and contamination can easily masquerade as a "microbiome." Careful work has challenged the notion of a stable placental microbiome in most pregnancies, even while acknowledging that pathogens can sometimes be present. For the infant's long-term microbial ecology, the moment of birth remains one of the clearest, highest-volume exposure events, and its consequences are amplified because the system is just beginning to assemble.

One way to make these dynamics intuitive is to think about what a pioneer species must accomplish. The newborn gut offers nutrients, but it also poses challenges: bile acids, an immune system on alert, fluctuating oxygen, and an evolving mucus layer. A successful pioneer must grow fast, tolerate stress, and interact with the host without provoking overwhelming inflammation. *Lactobacillus* species, when they appear early, bring an acidifying metabolism and a history of living on mucosal surfaces. *Staphylococcus* species, common on skin, are also tough pioneers, but they come with different ecological preferences and different interactions with host defenses. The early balance between these kinds of organisms can tilt the chemical environment, which then influences which strict anaerobes can expand.

The long-term consequences are still being disentangled. Some associations link C-section delivery to modestly increased risks of asthma, allergies, and certain autoimmune conditions, though these links are complicated by confounding factors (why a C-section was performed, maternal health, antibiotics, prematurity, and socioeconomic variables). Microbiome research has taught an uncomfortable lesson repeatedly: correlation is easier to find than causation is to prove. Yet the mechanistic plausibility here is stronger than in many other areas, because we can see the exposure difference, measure the microbial difference, and describe a developmental window in which the immune system is unusually plastic.

There is another practical implication that often gets lost in debates over delivery mode: the early microbiome is not shaped by one event alone. It is shaped by a cluster of linked events. A planned C-section in a stable pregnancy is not the same as an emergency C-section after prolonged labor, ruptured membranes, repeated vaginal exams, and maternal antibiotics. A vaginal birth with intrapartum antibiotics for Group B streptococcus prophylaxis is not the same as a vaginal birth without antibiotics. A preterm infant in a neonatal intensive care unit experiences a different microbial world from a term infant at home. When people ask whether delivery mode "determines" the microbiome, the honest answer is that it sets initial conditions in a sensitive period, and then the rest of early life either amplifies or dampens those initial differences.

The founder effect is easiest to see when the system is pushed. Consider an infant exposed to broad-spectrum antibiotics early in life. Antibiotics can suppress key early colonizers and open temporary space for opportunists, including drug-resistant strains. In the adult gut, such disturbances can sometimes lead to dramatic overgrowths, as in *Clostridioides difficile* infection after antibiotics. Infants are not simply small adults, but they share the ecological vulnerability: when a community is assembling, removing a large fraction of its early members can change the developmental path. Delivery mode is not antibiotics, but the logic is similar. Early disruptions can echo.

Because the differences are probabilistic rather than absolute, it helps to visualize delivery mode not as a fork in the road with two separate destinies, but as two different starting palettes. Many infants will converge toward broadly similar adult-like communities over time. Yet the first colors laid down can influence the blends that become possible later, especially when other pressures act on the system.

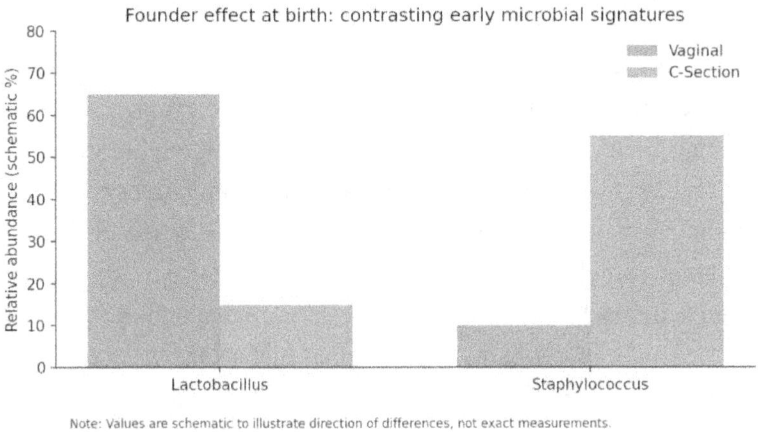

If this feels unsettlingly contingent, it should. Human biology is not only encoded in DNA; it is also scaffolded by relationships with microbes that we acquire through contact, chance, and care. Birth is one of the few moments when the microbial world can be redirected by a single clinical decision, not because clinicians are choosing microbes, but because they are choosing an environment. The operating theatre and the birth canal are both marvels of modern life in their own ways, each with its own microbial ecology, and the newborn arrives ready to be colonized by whatever world meets it first.

The infant's gut, meanwhile, behaves with the blunt pragmatism of ecosystems everywhere: it rewards whatever can grow under current conditions and it remembers, for a time, the consequences of that early growth. The first colonizers consume oxygen, reshape chemistry, and train immune expectations. Later arrivals do not enter an untouched habitat; they enter one already modified by the pioneers. In that sense, the landscape of the microbiome is not merely seeded at birth. It is landscaped.

3.2 Milk as Code

Milk looks, at first glance, like a straightforward answer to a straightforward problem: a new human needs calories, water, fats for a growing brain, and amino acids for rapidly multiplying cells. Yet mammalian milk contains ingredients that the infant cannot digest. This is not a minor trace oddity. In human milk, some of the most abundant carbohydrates are complex sugars that pass through the infant's stomach and small intestine largely untouched. From the point of view of the baby-as-an-engine, that seems wasteful. From the point of view of the baby-as-an-ecosystem, it makes perfect sense.

Those complex sugars are called human milk oligosaccharides, usually shortened to HMOs. An oligosaccharide is a short chain of sugar units, built into elaborate branching shapes. Lactose, the familiar milk sugar, is simple: it is readily split and absorbed by the infant. HMOs are different. They have many distinct structures, and they are decorated with chemical groups such as fucose and sialic acid. The infant's digestive enzymes are not well suited to breaking many of these bonds. So HMOs survive long enough to reach the lower intestine, where microbial communities are assembling and competing.

This is where the idea of milk as code becomes useful. Codes are not merely fuel; they are messages that shape behavior. In nature, a flower's scent is a code that recruits particular pollinators, and a bird's song is a code that recruits mates and repels rivals. Human milk, in this sense, does not only recruit the infant into survival. It recruits specific microbial partners into residence.

The relationship is deeply asymmetrical at first: the infant provides warm space and constant food; the microbes provide chemical labor and immune training. But that labor is not abstract. Many microbes can ferment simple sugars when given the chance. Far fewer can efficiently digest HMOs. The surprising result is that milk appears designed, in part, to feed not the infant directly, but a select group of microbes that, in turn, help protect the infant.

The historical clue that something like this was happening predates modern sequencing by many decades. In the early twentieth century, pediatricians and microbiologists noticed that breastfed infants often had stools dominated by bifidobacteria, giving rise to talk of a "bifidus flora." The notion of a "bifidus factor" in human milk circulated long before anyone knew the structures of HMOs. It was an inference from a pattern: breastfed babies tended to cultivate a particular microbial community, and formula-fed babies often did not. Only later did chemistry catch up to the observation. Lars Bode and others, writing in the early 2010s, helped crystallize the modern view: HMOs are not incidental; they are abundant, diverse, and plausibly adaptive.

A good way to appreciate what HMOs do is to imagine the infant gut as a newly opened market. Many would-be merchants show up, but only some can accept the currency. Lactose is a widely accepted currency; countless microbes can spend it. HMOs are a specialized currency, legible to a narrower set of readers. Among the best readers is *Bifidobacterium longum* subspecies *infantis*, often abbreviated *B. infantis*. This organism carries gene clusters that encode the transporters and enzymes needed to import and break down a wide array of HMO structures. In ecological terms, HMOs are a selective resource: they tilt the competition toward microbes that can harvest them efficiently.

That tilt matters because early-life microbes are not just passengers. They interact with the gut lining and with the immune system at a moment when both are learning their baseline settings. A microbiome dominated by bifidobacteria tends to produce metabolic byproducts, such as lactate and acetate, that lower pH and can make life harder for certain pathogens. It can also influence the integrity of the gut barrier, the single-cell-thick epithelium that must both absorb nutrients and keep microbes at arm's length. A leaky barrier in a newborn is not a small issue; it can invite inflammation at a time when inflammation carries outsized risks.

HMOs provide a particularly elegant demonstration of how the microbiome organ blurs boundaries between "self" and "non-self." The infant's genome does not encode the full ability to process HMOs. The mother's mammary glands synthesize them at energetic cost. The microbes encode the enzymes to digest them. The output is a three-party economy: mother invests, microbes metabolize, infant benefits.

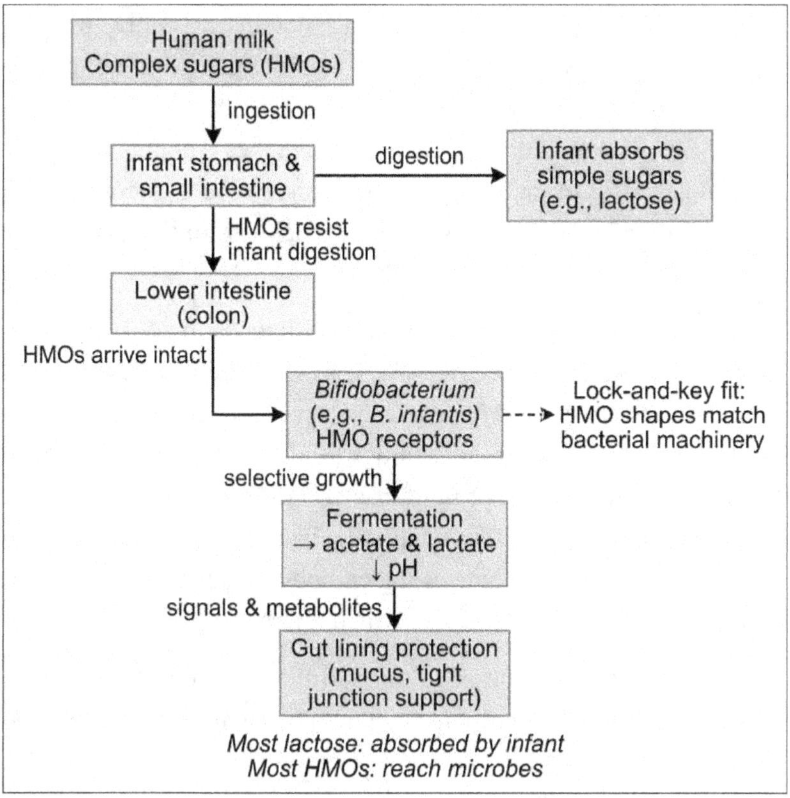

The phrase "lock-and-key" is often used in biochemistry for enzymes and receptors, and it fits here without much strain. HMOs are not just long strings; their shapes and decorations matter. Certain bifidobacteria have surface proteins and transport systems that recognize these shapes, bring them into the cell, and dismantle them step by step. The details vary by strain, and that variation is part of the point: milk does not simply favor "bifidobacteria" in general. It can favor particular kinds of bifidobacteria with the right equipment.

Milk, however, is not a single, uniform substance. It varies between mothers, over the course of lactation, and even within a day. One of the most striking sources of variation comes from maternal genetics, especially a gene called *FUT2* (fucosyltransferase 2). This gene helps determine "secretor status," a trait first noticed not in the context of milk, but in blood group antigens and their presence in bodily fluids. Secretor mothers tend to produce higher amounts of certain fucosylated HMOs, including a famous one called 2′-fucosyllactose (2′FL). Non-secretor mothers produce much less of these particular structures.

The microbial consequences of this difference have been documented with increasing clarity. Work by Zachery Lewis, Sarah Totten, and colleagues (2015) found that breastfed infants of secretor mothers were more likely to establish bifidobacteria earlier, and that the species composition of bifidobacteria differed depending on secretor status. This is milk as individualized ecological instruction: the mother's genotype changes the carbohydrate landscape, and the carbohydrate landscape changes which microbes flourish.

The practical implications do not require any genetic determinism. Secretor status does not doom or guarantee anything; it shifts probabilities, like so much in biology. But it underlines a larger idea: the mother is not only providing calories; she is shaping the infant's microbial niche with biochemical specificity.

To see why that might have been favored by evolution, consider the vulnerability of infants to gastrointestinal infection. A newborn does not have a mature immune repertoire. Its gut barrier is still developing. Inflammation, while often protective, can damage fragile tissue. A strategy that recruits microbes capable of lowering gut pH and occupying attachment sites, while also training the immune system toward tolerance rather than overreaction, would have been a powerful survival advantage in ancestral environments where clean water and sterile feeding equipment did not exist.

HMOs likely contribute to protection in more than one way. First, they act as prebiotics: nutrients for microbes that benefit the host. Second, many HMOs can act as soluble decoys. Pathogens often begin an infection by binding to specific sugar patterns on the surface of gut cells, a molecular handshake that lets them attach and invade. If milk contains free-floating sugars that mimic those surface patterns, the pathogen can bind the decoy instead, like a key fitting into a fake lock. It is a simple idea with an elegant chemistry: flood the gut with harmless look-alikes and make it harder for the wrong organisms to find purchase.

Third, HMOs and the microbes that consume them influence immune signaling. When bifidobacteria ferment HMOs, they produce short-chain fatty acids and other metabolites that can affect epithelial cells and immune cells. These chemical messages can help tighten the junctions between gut cells, encourage mucus production, and modulate inflammation. The immune system is not being "turned off." It is being educated: learning that certain microbial presences are normal and that responses should be proportionate.

Milk also carries immune components directly. Secretory IgA (SIgA) is an antibody abundant in human milk. Unlike antibodies that circulate in the blood, SIgA is specialized for mucosal surfaces. It can bind microbes in the gut lumen, sometimes preventing them from attaching to the gut wall, sometimes shaping their behavior more subtly. The striking detail is that SIgA is not merely a blunt antimicrobial smear. Evidence increasingly suggests specificity: SIgA can discriminate among microbes and can be shaped by maternal exposures.

Recent work by Katherine Donald, Antonio Serapio-Palacios, and colleagues (2025) highlighted how milk SIgA can act like a selective force, promoting colonization by some strains while restricting expansion of others, even when those strains are closely related. That kind of selectivity fits the idea of milk as code with almost unsettling precision. It suggests that breastfeeding can transmit not only generalized immune protection but a tuned set of signals that influence which microbial partners are welcomed.

These milk-derived instructions interact with the founder effect from birth. Early colonizers arrive first, but milk determines which of them can thrive. A microbe that reaches the gut but cannot exploit the available resources will be outcompeted. A microbe that can consume HMOs, tolerate bile, and avoid provoking damaging inflammation has a head start. Over days and weeks, the community begins to look less like a random sample of the environment and more like a system shaped by host selection.

A concrete example can make this feel less abstract. Imagine two infants with similar births and similar households. One is breastfed, the other receives a formula lacking HMOs (as was typical for many decades). In the breastfed infant, HMOs pass through and nourish HMO-specialists such as *B. infantis*. Those bifidobacteria produce acids that lower gut pH, and they tend to dominate. In the formula-fed infant, the available carbohydrates are more digestible by the infant, leaving fewer complex substrates reaching the colon, and a broader mix of microbes may share the remaining resources. The result is often a community with less bifidobacterial dominance and a different balance of metabolic outputs. This is not a moral fable about "natural" versus "artificial." It is an ecological fable about selective resources.

Modern formula has increasingly tried to close that gap by adding prebiotics and, in some regions and products, specific HMO structures. This is a fascinating public health experiment in real time: can one or a few HMOs reproduce the effects of the whole complex mixture? HMOs are not one molecule but a library, and the library's diversity may matter. So may the timing, the dose, and the presence of the right microbial strains to read the code. A code only works if the receiver has the right alphabet.

Timing is especially important because the infant gut does not stay in its early state. As solid foods appear, the carbohydrate landscape changes. Plant fibers and resistant starches begin to reach the colon, favoring different microbial guilds. At some point, breastfeeding typically decreases or stops, and that transition can be more consequential for the microbiome's maturation than the first spoonful of pureed carrot. Metagenomic work from Fredrik Bäckhed's group (2015), among others, has emphasized that cessation of breastfeeding is a major shift toward an adult-like microbiome. Milk, in other words, is not simply a starter culture; it is a sustaining influence.

One must be careful not to overpromise. The microbiome has a way of humbling grand stories. Not every breastfed infant develops the same community. Antibiotics can disrupt colonization. Prematurity changes gut physiology and hospital exposures. Maternal secretor status changes the HMO repertoire. Some infants may not carry *B. infantis* at all, depending on geography, household transmission, and broader social practices. And even when bifidobacteria dominate early, the long-term health implications are complex and often difficult to disentangle from confounders. Still, as a piece of evolutionary logic, milk-as-code is almost too neat to ignore: humans spend substantial metabolic energy producing molecules that largely bypass infant digestion, and those molecules reliably feed microbes that tend to protect the infant.

There is also something emotionally resonant, even a little unsettling, in the notion that breastfeeding is partly aimed at feeding someone else. The infant is not the only mouth at the table. But that framing can be made gentler: the infant is not outsourcing nutrition, but recruiting allies. A newborn cannot digest plant fiber, cannot produce adult levels of antimicrobial peptides, and cannot mount mature immune responses without risk. Instead, it enters into a managed partnership. Milk provides the substrates and signals that make the partnership more likely to be beneficial than dangerous.

The most revealing way to think about HMOs may be to treat them as invitations with strict dress codes. Many microbes will show up; few will be admitted. The ones that gain entry help renovate the gut into a safer home: a little more acidic, a little less permeable, and populated by residents that the immune system learns to tolerate without going to war. In the small hours when a baby feeds half-asleep, the transaction feels intimate and immediate: comfort, warmth, the quiet work of growing. Underneath, a biochemical letter is being delivered to the gut, addressed not only to the infant's cells but to the microbes that can read it, asking them to settle in and, in exchange for room and board, to help keep the peace.

3.3 The Long Fade

The gut microbiome is often described with the language of childhood: it is "established," "trained," "matured." That vocabulary flatters adulthood as a stable endpoint, as if the microbial community inside us simply locks into place and then sits quietly, like a well-run library. For many people, something like that is true for a while. The adult gut tends to be more consistent from month to month than the infant gut, and it can resist small disturbances. But stability is not permanence. In later life, the microbiome begins to change again, and the pattern is less like a tidy maturation than a slow unravelling.

Aging is not a single process; it is a collection of changes that arrive on different schedules. Teeth and appetite change. Mobility changes. Sleep changes. Medications multiply. The immune system shifts from agile responsiveness toward a chronic, low-grade inflammatory state that gerontologists often call *inflammaging*. The gut, too, changes: motility can slow, the mucus layer can alter, and the barrier that keeps microbes at the proper distance can become more permeable. Each of these factors nudges the microbiome. Together they can produce a striking ecological outcome: loss of diversity.

In ecology, diversity can mean several things, but the simplest sense is "how many different kinds are present, and how evenly they share the space." A diverse forest has not only many species, but also a structure that makes it resilient: if one tree species suffers a blight, others can fill the gaps. The gut is not a forest, yet it faces similar problems. A community with many functional capabilities can keep digestion and immune signaling on a steady course even when the diet changes or a virus arrives. When diversity declines, the community becomes simpler, often more dominated by a smaller number of organisms. That simplification can make the system more fragile. It can also change what chemical products are made in the gut, because microbes are chemical factories, and fewer factories mean fewer outputs.

The association between microbial diversity and healthy aging has been sharpened by studies that looked not only at age, but at the lived realities that come with age: diet, frailty, and residence setting. One influential example came from ELDERMET, a large Irish project that profiled the gut microbiota of older adults living in different contexts. The striking observation was not merely that older people had different microbes than younger people, but that older adults in long-term residential care often had lower microbial diversity than community-dwelling adults, and that this shift correlated with measures of frailty and markers of inflammation. The microbiome seemed to track a particular kind of aging: not chronological age in years, but biological and social aging, the kind that shows up as weakness, weight loss, and vulnerability to infection.

Frailty is a clinical term with a plain meaning. It describes a state in which the body has less reserve. Small stressors that a robust person would shrug off can cause disproportionate harm. A minor infection can lead to delirium; a short hospital stay can lead to lasting weakness; a brief appetite loss can tip into malnutrition. Frailty is not identical to disability, and it is not inevitable, but it is a real and measurable syndrome.

Why would microbial diversity relate to frailty? One answer lies in diet. Many of the microbes associated with a diverse, "community-like" gut are specialists in fermenting complex plant fibers into short-chain fatty acids, such as butyrate. Butyrate is more than a waste product. It is a primary fuel for colon cells and a molecule with anti-inflammatory effects in several experimental settings. Diets rich in varied plant fibers tend to support microbes that produce these metabolites. In older age, diets often narrow, sometimes for reasons that have nothing to do with preference. Dental problems make raw vegetables difficult. Reduced smell and taste flatten the pleasure of eating. Living alone can reduce the motivation to cook. Institutional food is designed for safety, cost, and consistency, not for microbial diversity.

Even when an older adult wants to eat well, illness can steal appetite. Depression can do the same. So can medications. A microbiome that is deprived of its usual substrates does not politely wait. It shifts. Fiber-fermenters decline, and organisms that can thrive on simpler inputs, on mucus-derived sugars, or on inflammatory byproducts may expand. The community becomes less diverse and often less aligned with the host's needs.

Another answer lies in the gut barrier. The gut lining is not only a wall; it is a negotiation. It must allow nutrients through while keeping microbes at a safe distance. With age and illness, the barrier can become leaky. A little more microbial material crosses into the body, and the immune system responds. The response is often not a sharp feverish defense, but a simmering inflammation that becomes part of the baseline. *Inflammaging* is not necessarily caused by the microbiome, but the microbiome can feed it, and *inflammaging* can, in turn, reshape the microbiome. In ecology, feedback loops matter. They can lock a system into a new state.

A concrete example makes the feedback loop easier to see. Consider an older adult who has been living independently with a reasonably varied diet. A fall leads to a hip fracture. Surgery follows, and with it pain medications, antibiotics, reduced mobility, and a hospital stay. Food becomes unfamiliar, and appetite is low. After discharge, rehabilitation may require a period in a nursing home. In that setting, exposure patterns change: new microbes circulate, shared surfaces multiply, and infection-control measures alter which organisms are most successful. Over weeks and months, the microbiome can drift away from its previous community pattern toward one shaped by institutional life and polypharmacy (the use of multiple medications). The person becomes weaker and less active. The feedback tightens: less activity and less dietary variety reduce the substrates that support a diverse microbiome; reduced diversity and altered microbial metabolism can contribute to gut dysfunction, inflammation, and reduced nutrient absorption; those physiological effects can worsen frailty.

This is not a morality tale about nursing homes. Long-term care facilities provide essential support and safety. The point is that residence setting is a powerful ecological variable. It changes both exposures and routines. In studies of nursing home residents, microbiome composition has been strongly associated not only with age and frailty but with medication exposures, including psychoactive drugs and antihypertensives. The gut microbiome is sensitive to chemicals, and in older adults, chemicals are often unavoidable. Each drug is prescribed for a reason, yet each can have side effects that ripple through the ecosystem: changes in motility, changes in saliva, changes in appetite, changes in gastric acidity, direct antimicrobial effects, and altered bile acid pools.

Time also matters. Some research has suggested a time-dependent pattern in long-stay care: during the first period of residence, the microbiome can become unstable and then, after a prolonged stay, regain a kind of stability around an altered state, one enriched for organisms associated with inflammation and depleted for organisms often considered symbiotic. Stability is not always good news. A stable swamp can be as stable as a stable meadow.

Historical anecdotes in microbiome science are often about discovery, but in aging the most revealing history is social. For most of human existence, many older adults lived in multigenerational households with shared meals, gardens, and routine exposure to outdoor environments. That world was not idyllic; it carried its own risks. Yet it did provide a particular kind of microbial continuity: shared kitchens, shared microbes, and daily movement. Modern longevity has been bought, in part, by medical triumphs and sanitation, but it has also produced new patterns of isolation. More people live alone for longer. More people spend their final years in institutions. Those changes alter microbial ecology as surely as they alter mental health.

There is a caution here, because the temptation is to treat microbial diversity as a simple health score, like blood pressure. More diversity, better health; less diversity, worse health. That heuristic has some truth, but it is not a law. Some infections raise "diversity" by adding transient species, and some very healthy infant guts are not especially diverse because they are dominated by bifidobacteria. Context matters. Function matters. An older adult could have a diverse microbiome that nonetheless lacks key fiber fermenters, and another could have lower diversity but a community rich in beneficial functions. Diversity is a clue, not a verdict.

Even so, the repeated association between loss of community-associated microbiota and frailty is hard to dismiss because it aligns with mechanisms that make biological sense. A less diverse community tends to produce fewer kinds of metabolites. If the microbes that convert fiber into butyrate decline, the gut lining may be less supported. If the microbes that metabolize bile acids shift, bile acid profiles may become more inflammatory. If opportunistic organisms expand, they may provoke immune activation. The overall effect can be a body in a mild state of siege, spending energy on managing inflammation and less on rebuilding muscle, maintaining bone density, and supporting cognition.

Muscle is an especially important part of this story. Sarcopenia, the age-related loss of muscle mass and strength, is a major contributor to frailty. Muscle is metabolically active tissue. It is also a reservoir of protein, a buffer against illness. When muscle declines, reserve declines. The gut microbiome can influence muscle indirectly through inflammation, appetite, and nutrient absorption, and possibly through specific metabolites that affect insulin sensitivity and energy balance. The evidence here is still developing, and it is easy to overstate. Yet it is not far-fetched that an ecosystem in the gut could influence an organ system as widespread as muscle, because the gut is a major site of immune activity and a major source of circulating microbial metabolites.

One of the most unsettling aspects of the long fade is that it can be quiet. People often imagine microbial problems as acute: food poisoning, diarrhea, a sudden infection. But the microbiome's contribution to frailty is more like slow corrosion. A diet becomes narrower; the microbiome shifts; low-grade inflammation rises; appetite drops; recovery from minor setbacks slows. Each step is modest. The sum can be large.

Not all changes are inevitable. Older adults who maintain varied diets, physical activity, and social engagement often retain microbiomes that look more "community-like." That observation itself can be emotionally complicated, because it risks sounding like advice delivered to people who may have limited control over their circumstances. It is worth saying plainly that frailty is not simply a lifestyle choice; it is shaped by economics, caregiving, housing, disability, and disease. Microbiome science does not erase those realities. It can, however, give them a biological dimension that is measurable and, potentially, modifiable.

The most direct modifiable factor is diet, especially dietary fiber. Fiber is not one thing; it is a category that includes many structures found in plants, each feeding different microbial specialists. A varied plant-based diet supplies a varied microbial substrate. In older adults, increasing fiber is not always straightforward, because gut motility may be slower, and sudden increases can cause discomfort. Yet gradual changes, tailored to the person, can matter. The goal is not to chase an abstract "diversity" number but to restore ecological functions: fermentation of complex carbohydrates, production of barrier-supporting metabolites, and suppression of opportunistic blooms.

Medications are another modifiable factor, though they must be handled with care. Deprescribing, the deliberate reduction of unnecessary medications, has become an important principle in geriatric medicine. When medications can be safely reduced, the microbiome may benefit indirectly through improved appetite, motility, and sleep. Antibiotics deserve particular attention. They are lifesaving, but they also exert a strong selective pressure on gut communities. In older adults, repeated antibiotic courses can push the microbiome toward low-diversity states and increase vulnerability to pathogens such as *Clostridioides difficile*, a bacterium that can exploit disrupted communities and cause severe colitis. The existence of fecal microbiota transplantation as an effective therapy for recurrent *C. difficile* infection is, among other things, a vivid demonstration that restoring community structure can restore health. It is a rare case where microbiome causation is not merely hypothesized but clinically enacted.

There is also a more subtle kind of intervention: preserving the rhythms that keep ecosystems stable. Regular meal timing, modest physical activity, exposure to daylight, and social meals can influence appetite and gut motility. These are not "microbiome hacks"; they are ordinary human patterns that happen to support ecological regularity. The gut is not only influenced by what enters it, but by how it moves and by the hormones and nerves that regulate it. Loneliness can change stress hormones, sleep, and inflammatory tone, and these can feed back into the gut environment. A microbiome is not an isolated jar of bacteria; it is a living organ intertwined with the nervous system and with the social world.

Aging, then, is not simply the accumulation of years in the host. It is also the gradual reshaping of the internal ecosystem: the resources available, the physical habitat, the immune climate, and the disturbances imposed by modern medical care. Diversity loss is one visible signature of this reshaping, correlated with frailty because both reflect reduced reserve. The cruel irony is that the microbial community that once helped build the host's resilience can, under the pressures of late life, lose the complexity that made it resilient in the first place.

Yet even in the long fade, the microbiome remains responsive. It is an ecosystem, not a fossil. It changes with meals, with medications, with environments, with companionship, with the small daily circumstances that determine whether an older person eats well, sleeps well, and moves without fear. The microbes respond to those circumstances with their own quiet arithmetic: growth rates, competition, chemical conversions. The sums of that arithmetic do not announce themselves, but they can be felt in the difference between recovery and decline, between a body that bounces back and one that cannot quite find its footing again.

Chapter 4

The Chemical Language

Our cells and our microbial residents converse not with words, but with a precise vocabulary of molecules that bridges the species gap. By transforming dietary fibers, digestive fluids, and amino acids into potent signals, the microbiome instructs our bodies on how to regulate inflammation, metabolism, and even neural states.

4.1 The Alchemy of Fiber

A paradox sits at the center of human nutrition. Some of the most important food we eat is, by definition, food we cannot eat.

Dietary fiber is made of carbohydrates, but not the kind that human enzymes can readily take apart. We can snip starch into glucose and absorb it; we can dismantle many sugars and drink their sweetness straight into the bloodstream. Fiber is different. It is the tough scaffolding of plants: chains of sugars locked together in bonds our own digestive chemistry does not recognize. So fiber passes through the small intestine almost untouched, like a sealed parcel addressed not to us but to the dense microbial metropolis living farther downstream.

If that sounds like wasted effort, it helps to picture the colon not as a mere exit corridor but as a working organ whose daily business is maintenance. It has to keep a barrier intact across a surface large enough to matter, under constant chemical and mechanical stress, while hosting trillions of microbes at arm's length. Its lining is renewed quickly; its immune system is vigilant without being trigger-happy; its mucus layer is manufactured like a protective gel-coat. None of this comes free. The colon needs energy and instructions, and much of both arrives through a route that bypasses our own digestion: fermentation.

Fermentation is a word with old, domestic associations. The Latin root *fervere* means "to boil," and early observers were not wrong to think something lively was happening when grape juice frothed into wine or cabbage soured into sauerkraut. In the nineteenth century, Louis Pasteur helped turn that kitchen mystery into a biological idea: microbes could transform food without oxygen, pulling energy from chemical rearrangements that looked, to the human eye, like spoilage or magic. In the gut, fermentation is not a culinary trick but a steady industrial process, running day and night, converting the plant parts we cannot digest into molecules our bodies can use and interpret.

The microbes that do this are not simply "eating fiber." They are performing an ecological relay. Large, knotted carbohydrates must be cut into smaller pieces first; different species specialize in different cuts. Some excel at breaking open long chains; others thrive on the fragments. The final products are small acids that, in the right context, become both fuel and message. These are the short-chain fatty acids, usually abbreviated to SCFAs: acetate, propionate, and butyrate.

Their names can sound like lab inventory, yet their effects are intimate. If you have ever felt the difference between a gut that seems calm and one that seems inflamed and jumpy, you have felt the downstream consequences of this chemistry. The colon's cells, the immune system's patrols, and the nerves that set the rhythm of the bowel all respond to SCFAs. In the right amounts and places, they are less like waste products and more like a shared language between microbes and host.

Before explaining why, it helps to clarify what fiber actually is. “Fiber” is a category, not a molecule. It includes soluble fibers that dissolve and gel (such as pectins and some beta-glucans), insoluble fibers that bulk and resist breakdown (such as cellulose), and a variety of resistant starches that behave like fiber because they escape digestion. Some fibers are easy for microbes to ferment; others are stubborn and pass through more intact. The mix matters. A gut community adapts to the substrates it regularly receives, and different fibers tend to favor different microbial talents. That means two people can eat “more fiber” and produce very different SCFA patterns, not because the fibers are identical, but because the microbial machinery reading the parcel differs from one colon to the next.

Once fiber reaches the colon, fermentation begins with enzymes that humans lack: microbial glycosidases and polysaccharide lyases, molecular scissors and crowbars. These break long carbohydrate chains into smaller sugars and oligosaccharides, which microbes import and metabolize. In the absence of oxygen, the microbial cell has to balance its internal books: it extracts energy while recycling electron carriers that would otherwise get stuck in the wrong chemical state. One way to do that is to dump certain end-products into the environment. The colon becomes a shared fermentation vat, and the end-products diffuse outward, bumping into host cells.

Acetate tends to be the most abundant SCFA. It is produced by many microbes and can travel beyond the gut into the bloodstream. Propionate is often associated with certain bacterial groups that channel fermentation toward three-carbon products; it too can reach the liver and participate in metabolic regulation. Butyrate is often the star of this story, not because it is the most plentiful but because it behaves like a local currency that the colon spends immediately.

Colonocytes, the cells that line the colon, can oxidize butyrate as a primary fuel. This fact alone changes the emotional tone of fiber. What looks like "indigestible" food becomes, through microbial labor, a direct energy source for an organ that must constantly renew itself. When butyrate supply is robust, colonocytes can meet much of their energetic demand in place, and that has consequences for barrier health: energy supports tight junctions between cells, mucus production, and the repair work that follows everyday friction. When butyrate supply dwindles, the system becomes more precarious. The colon is then pushed to rely more on other fuels, and the ecological balance can shift in ways that erode the calm separation between microbes and host tissue.

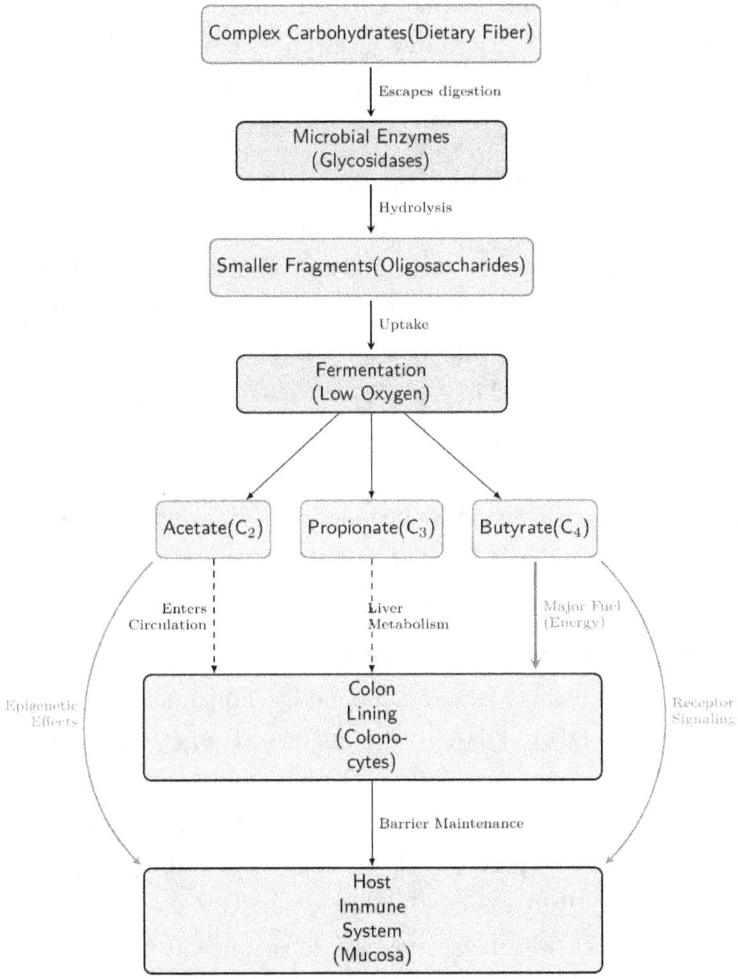

Energy is only half the story. The other half is interpretation.

The immune system in the gut lives with a dilemma. It must be tolerant of food and friendly microbes, yet ready to act against pathogens and breaches. That tension is not abstract; it is written into the architecture of the gut wall. Immune cells sit beneath a single layer of epithelial cells. A thin sheet of tissue, not much thicker than a coat of paint, separates the internal body from the microbial world. That thinness is a strength (it allows absorption and surveillance) and a vulnerability (it is easy to disrupt). Keeping peace requires constant negotiation.

SCFAs are part of that negotiation because they operate as signals as well as nutrients. Cells of the host possess receptors that detect SCFAs, particularly on immune and epithelial cells. These receptors are a kind of molecular nose: when acetate, propionate, or butyrate binds, the cell changes what it does. The details vary by receptor and cell type, but the broad idea is simple enough for a dinner party: the host has evolved to treat microbial fermentation products as a readout of ecological conditions. When fiber is plentiful and fermentation is steady, SCFAs rise, and immune tone shifts toward maintenance and tolerance. When the SCFA pattern changes, it can be a clue that the microbial economy is shifting.

Butyrate, in particular, has a second mode of action that feels almost poetic: it can change gene expression by influencing the chemical packaging of DNA. DNA in the nucleus is wrapped around proteins called histones, and cells adjust how tightly DNA is wrapped to control which genes are accessible. Enzymes known as histone deacetylases (HDACs) help tighten this packaging by removing acetyl groups. Butyrate can inhibit some HDACs. That inhibition tends to increase histone acetylation, loosening chromatin and making certain genes easier to turn on.

This is not mind control by bacteria; it is chemistry. Yet the consequences can be profound. If an immune cell receives signals that favor regulation over aggression, and if butyrate shifts gene accessibility in a way that supports that regulatory program, the immune response in the gut can become less prone to hair-trigger inflammation. One of the best-known examples involves regulatory T cells, often shortened to Tregs, which help prevent immune responses from overshooting and damaging tissue. Experiments in mice in the early 2010s linked butyrate to increased Treg differentiation in the colon, showing a plausible route from diet to microbial metabolism to immune training.

That word *training* matters because the gut's immune system is not a static guard. It is educated by experience. From infancy onward, exposure to microbes and their metabolites shapes how immune cells respond. Fiber, then, is not only about bowel habits or cholesterol numbers. It can act as a kind of curriculum, delivered in chemical form, with microbes as the teachers and SCFAs as the lesson plan. The content of the curriculum depends on what the microbes can make, and that depends on what we feed them.

This is where ecological nuance enters. It is tempting to hear "butyrate is anti-inflammatory" and file it away as a universal truth, like vitamin C prevents scurvy. Biology rarely behaves so cleanly. SCFAs influence immune cells within a broader context of cytokines (immune signaling proteins), microbial components, host genetics, and the integrity of the barrier itself. In some circumstances, the same SCFA signals can support effector functions, not just regulatory ones. A cell's interpretation depends on what other signals are present. The immune system is more like a parliament than a single ruler: votes are counted, and outcomes depend on coalitions.

Still, the association between fiber fermentation and gut calm is strong enough to shape clinical thinking. Consider an example that shows how the microbial ecosystem can be destabilized and what that does to the gut wall: antibiotic-associated diarrhea and *Clostridioides difficile* infection. Antibiotics can be lifesaving, but they often act like a wildfire through the microbiome, reducing diversity and removing key organisms that keep the ecosystem stable. After such a disruption, *C. difficile* can seize the opportunity, producing toxins that injure the colonic lining and provoke severe inflammation.

The story is not "low fiber causes *C. difficile*." The story is that the gut's colonization resistance depends on a resilient community and a well-supported barrier. SCFAs are one of the currencies in that resilience. A community that efficiently ferments fiber tends to acidify the environment, feed colonocytes, and support mucus and barrier function. When the community is flattened, those services can falter, and opportunists find room to expand.

If this makes fiber sound like a simple prescription, it is worth pausing to respect how personal the chemistry can be. Two people can eat the same bowl of oats and beans and generate different SCFA profiles. Part of that is microbial composition. Part of it is where fermentation happens: some fibers are fermented earlier in the colon, others later. Location matters because different parts of the colon face different challenges. The distal colon, closer to the exit, may have less fermentable substrate left, and the microbes there may shift to alternative fuels, including mucus-derived sugars if dietary fiber is scarce. That shift can thin the mucus layer, bringing microbes closer to the epithelium and increasing the chance of low-grade inflammation. A fiber pattern that delivers fermentable substrate deeper into the colon can be especially valuable for maintaining that distal barrier.

The idea of microbial cooperation also matters, because butyrate is often produced through cross-feeding. One set of microbes breaks down complex carbohydrates into simpler products, including acetate and lactate; another set converts those into butyrate. This division of labor means that butyrate production can collapse if a key link in the chain is missing, even if other microbes remain. It also means that adding a single "good" microbe is not always enough; the ecological context has to support its job. This helps explain why probiotic effects can be inconsistent and why dietary patterns that support a broader network of fermenters often have more durable effects than any single supplement.

A concrete way to picture this is to imagine a crowded apartment building where residents share utilities. One group collects recyclables; another runs a workshop turning them into something useful. If the collectors leave, the workshop starves; if the workshop closes, the recyclables pile up. In the colon, acetate can be both an end-product and a feedstock. When acetate is efficiently converted into butyrate by specialist microbes, the net output shifts toward a metabolite that colonocytes can burn. When that conversion fails, the system may still produce SCFAs, but in a different pattern, with different biological consequences.

What does this mean for everyday life? It suggests that the benefits of fiber are partly immediate and partly slow. Immediate, because fermentation products appear within hours and influence gut motility, water handling, and local metabolism. Slow, because microbial communities adapt to habitual diet. A person who suddenly doubles fiber intake may experience gas and discomfort at first, not as a sign of harm but as evidence that fermentation has sped up faster than the ecosystem and the gut's sensory system can comfortably accommodate. Over days to weeks, microbial gene expression and community structure can shift, often making the process smoother. The gut is not a passive tube; it is a responsive habitat.

It also clarifies why fiber can feel like it affects the whole body. Acetate and propionate can circulate, reaching organs beyond the gut. They can influence appetite regulation, lipid metabolism, and glucose handling through a mixture of direct signaling and indirect hormonal effects. But even when the action is local, local is not trivial. A healthier barrier means fewer inflammatory sparks entering the bloodstream. Chronic, low-grade inflammation is a theme in many modern diseases, and the gut barrier is one of the places where such inflammation can be kindled or quenched.

There is an older historical echo here, one that predates modern microbiome science. Physicians long noticed that diets rich in coarse plant foods seemed to protect against constipation and certain bowel ailments. In the 1970s, the surgeon Denis Burkitt, working in Africa and later comparing disease patterns across populations, argued that "Western" low-fiber diets were linked to a range of conditions, from diverticular disease to colon cancer. His claims were broader than what evidence can cleanly support, and later research refined and sometimes challenged the details. Yet Burkitt helped reframe fiber as more than roughage: it was a missing ingredient in a diet that had changed faster than human physiology. Today, SCFAs offer a mechanistic bridge between those epidemiological observations and molecular biology. They provide a plausible reason why the colon might suffer when deprived of fermentable substrate, and why restoring that substrate can support the gut's architecture and immune diplomacy.

Even so, caution is part of honesty. Most microbiome research in humans still struggles with the problem of causation: do SCFAs and their producing microbes *cause* better gut health, or do healthier guts simply support communities that make more SCFAs? Animal models and mechanistic studies make causality plausible, but human bodies live in a tangle of variables: medications, stress, sleep, infections, genetics, and countless dietary details. Measuring SCFAs is itself tricky. Levels in stool are not the same as production rates, because colonocytes can absorb and consume SCFAs before they ever appear in feces. A person can have low stool butyrate because little is being made, or because it is being efficiently used. The biology does not yield its secrets easily.

Yet the basic alchemy remains compelling: fiber becomes something else, and that something else matters.

One way to feel the logic in your bones is to remember that the colon evolved to host fermentation. Many mammals rely heavily on microbial fermentation for energy; in ruminants like cows, fermentation happens in a specialized foregut chamber and feeds the animal. Humans are not ruminants, but we are not fermentation-free either. Our colon is a fermentation organ attached to a primate body, and modern diets often feed it poorly. When we supply fermentable substrates, we are not indulging a fad; we are participating in an older arrangement between host and microbes.

The arrangement is not sentimental. Microbes ferment fiber because it benefits them, and the host benefits because it can use the products. Peace holds when both sides gain more than they lose. SCFAs help keep that peace by feeding the cells that form the wall and by nudging immune cells toward restraint when restraint is wise. They are not a magic shield against all inflammation, nor a guarantee of health. They are, rather, part of the gut's internal economy: energy that doubles as information, minted from the parts of plants our own enzymes cannot touch.

When that economy is functioning well, the colon is not simply transporting leftovers. It is quietly renewing itself, tightening its seals, producing its mucus, and instructing immune cells to treat the resident microbial crowd as neighbors rather than invaders. Fiber, in this sense, is not the absence of calories. It is the raw material for a conversation that keeps the boundary between self and non-self both firm and flexible, strong enough to protect and relaxed enough to live.

4.2 The Bile Refinery

Bile begins as a blunt tool. The liver makes it from cholesterol, packages it with salts and pigments, and sends it into the intestine when a meal arrives. There is nothing subtle about its first job: it behaves like detergent, breaking up dietary fat into tiny droplets that enzymes can attack. Without bile, a buttery dinner turns into an absorption problem, and the body's relationship to fat-soluble vitamins (A, D, E, and K) quickly becomes precarious.

Detergents, though, are rarely polite. They disrupt membranes; they dissolve grease by doing chemistry that cells would rather not do to themselves. For that reason, bile acids—the active detergents within bile—have to be handled carefully. The liver modifies them before release, usually by attaching amino acids that make them more soluble and less likely to seep into cells in the wrong place. The gallbladder stores this mixture, then delivers it into the small intestine in timed bursts. Most bile acids are later reabsorbed and returned to the liver, a recycling loop called the enterohepatic circulation. It is thrifty physiology: bile acids are expensive to make, so the body reuses them.

That recycling loop creates a predictable rhythm: bile acids flow out, travel down the gut, and then are pulled back into the body. The surprise is what happens along the way. Between departure and return, bile acids pass through a zone where human control is partial at best. In the lower small intestine and colon they meet dense microbial communities, and those microbes treat bile acids not as finished products but as raw material. They cut, twist, and remodel them. By the time many bile acids come back to the liver, they are no longer exactly what the liver sent out.

Calling this a "refinery" is not just a metaphor. In an oil refinery, crude input is separated and chemically modified into products with different properties and uses. In the gut, the liver provides the crude input and microbes generate a diverse output: new bile acids with altered detergent strength, different absorption patterns, and, crucially, different signaling effects across the body. A substance whose job began as dish soap ends up behaving like a hormone.

A hormone, in everyday terms, is a chemical message released in one place that affects cells elsewhere. Bile acids fit the definition because the body has receptors-molecular locks-built to be opened by particular bile acids. Two of the best-studied are FXR (a receptor that sits in the cell nucleus and controls gene expression) and TGR5 (a receptor on cell membranes that triggers fast signaling inside the cell). These receptors are present not only in the intestine but also in the liver, fat tissue, immune cells, and other organs. When bile acids bind them, they can influence glucose metabolism, lipid handling, energy expenditure, and inflammatory tone. The gut, in other words, is not simply a tube where bile does its grease-cutting and exits. It is a switchboard where bile acids are rewritten into messages.

The rewriting begins with chemistry that sounds small but has large consequences: deconjugation. The liver often attaches glycine or taurine to bile acids before secretion. This makes bile acids more water-soluble and helps keep them in the gut's watery environment. Many gut bacteria make enzymes called bile salt hydrolases (BSHs) that remove these attachments. It is a single snip, like unhooking a keychain, yet it changes how bile acids behave: their solubility shifts; their ability to be absorbed changes; their susceptibility to further microbial transformations increases.

Deconjugation also changes ecology. Bile acids are antimicrobial at high concentrations; they can damage bacterial membranes. Microbes that thrive in the small intestine often carry strategies to resist bile, and BSH enzymes are part of that. By deconjugating bile acids, bacteria can reduce the toxicity of the bile acid pool to themselves and their neighbors, adjusting the chemical climate of the gut. This is not altruism; it is survival. But the host experiences it as a reshaping of the bile acid landscape.

After deconjugation comes a step with an ominous name: 7α-dehydroxylation. This reaction removes a hydroxyl group from the steroid-like scaffold of certain bile acids. It is not common across gut microbes; it is concentrated in relatively few anaerobic species. Yet its impact is out of proportion to its rarity because it generates some of the most famous "secondary bile acids." Primary bile acids are those the liver makes directly. When microbes perform transformations like 7α-dehydroxylation, they produce secondary bile acids such as deoxycholic acid and lithocholic acid.

The labels *primary* and *secondary* can sound like a simple hierarchy, but the reality is messier, more like a family tree with many branches and cousins. Beyond the classic secondary bile acids, microbes can oxidize, reduce, and epimerize bile acids-subtle changes that flip the orientation of chemical groups and create new variants. In the past decade, microbial genetics has made it clear that some of these variants arise from dedicated biosynthetic pathways, not just opportunistic tinkering. Different people, carrying different microbial genes, can have very different bile acid profiles, even on similar diets. That means the hormonal messages carried by bile acids are partly personalized: the same liver input can be refined into different chemical outputs.

Why would the body tolerate, or even welcome, this microbial meddling? One answer is that it has little choice: the gut is colonized, and bile acids must pass through. But evolution tends to turn inevitability into opportunity. If microbial transformations are stable features of mammalian life, then host physiology can evolve to interpret them as signals. The existence of receptors like FXR and TGR5, and their sensitivity to different bile acids, makes sense in that light. The host is not only defending itself against microbial chemistry; it is reading it.

Consider FXR. In the intestine, FXR activation can influence the production of a hormone-like factor called FGF19 in humans (FGF15 in mice), which travels to the liver and helps regulate bile acid synthesis. This creates a feedback loop: bile acids activate FXR; FXR triggers signals that tell the liver how much bile acid to make. Microbial modifications that change which bile acids are present can therefore shift the feedback. The loop is elegant, but it is also a point of vulnerability: if the bile acid pool changes dramatically, the liver's regulation changes with it.

TGR5 offers a different style of influence. It is expressed in multiple tissues, including enteroendocrine cells in the gut that release hormones affecting appetite and glucose control. When certain bile acids activate TGR5, they can promote secretion of GLP-1, an incretin hormone that boosts insulin release after meals and slows gastric emptying. This is one reason bile acids have become relevant to metabolic disease. It also hints at why some interventions-bariatric surgery is a prominent example-can have rapid effects on glucose metabolism that outpace weight loss. Changes in gut anatomy alter bile flow and microbial ecology, which can shift bile acid signaling and gut hormone release. The details are still being worked out, and claims can outrun evidence, but the broad picture is hard to ignore: bile acids connect the gut to metabolism with surprising speed.

A historical thread helps here. In the nineteenth century, when bile pigments and acids were first being characterized, bile was treated mainly as a digestive secretion and a clinical clue: jaundice signaled liver disease, pale stools suggested blocked bile flow. The idea that bile acids might act as hormones would have sounded odd, even fanciful. Endocrinology was still young; receptors were not yet part of biology's common vocabulary. The conceptual shift came slowly, as molecular biology made it possible to identify receptors and trace their downstream effects. When receptors like FXR were characterized in the late twentieth century, bile acids began to look less like soap and more like information.

Microbes add another layer to that conceptual shift. If bile acids are signals, then microbial transformations are edits to the message.

That editing can be helpful or harmful depending on context. Some secondary bile acids, for example, are more hydrophobic and can be more irritating to tissues at high concentrations, implicated in colon carcinogenesis in some models and epidemiological associations. At the same time, certain secondary bile acids can suppress pathogens and shape immune responses in ways that support intestinal stability. Biology rarely hands us a single moral.

A concrete example of the ecological power of bile acids comes from *Clostridioides difficile*, the same opportunist that can bloom after antibiotics. *C. difficile* spores germinate in response to certain bile acids, and its growth can be restricted by others. Antibiotics can disrupt microbial species that normally transform bile acids, shifting the balance toward forms that encourage spore germination and away from forms that inhibit outgrowth. This creates a chemical welcome mat for infection. In mouse studies, restoring microbes with the right bile-transforming capacity, or restoring key enzymatic steps, can help re-establish resistance. Here, bile acids are not merely digestion aids; they are gatekeepers of colonization.

Deconjugation by BSH enzymes is particularly interesting in this context because it sits near the beginning of the microbial bile-acid assembly line. If conjugated bile acids remain conjugated, many downstream transformations are less efficient or altered. BSH activity can therefore act like a bottleneck that determines which bile acids are available for further refinement. Different BSH enzymes also show preferences: some are more effective on taurine-conjugated bile acids, others on glycine-conjugated ones. Small structural differences in the enzyme's active site can change its substrate range. That is a molecular way of saying something that is increasingly relevant in microbiome science: it is not enough to know which bacterial species are present; it matters which genes and enzymes they carry.

The refinement of bile acids also feeds back on the microbes themselves. Bile acids can suppress some bacteria and encourage others. A microbial community that changes bile acids changes its own environment, selecting for bile-tolerant species and shaping the competitive landscape. This creates the possibility of self-reinforcing states. For example, a high-fat diet increases bile secretion. Increased bile can select for microbes that tolerate bile well. Those microbes may carry enzymes that shift bile acids toward particular forms, which then further tune the environment. The host is not outside this loop; it is one of the players. The liver adjusts bile synthesis, the gut adjusts absorption, and the immune system responds to the changing chemical and microbial milieu.

This is where bile acids become a plausible link between diet, microbiota, and systemic disease-plausible, not proven in every detail. In obesity and type 2 diabetes, bile acid profiles often differ from those in lean, metabolically healthy individuals. Whether those differences are drivers, passengers, or both remains under debate. Yet mechanisms exist that could make them drivers: altered bile acids can change FXR and TGR5 signaling; those changes can influence hepatic glucose output, insulin sensitivity, GLP-1 secretion, and energy expenditure. Meanwhile, bile acids can affect inflammation, both in the gut and potentially systemically, by acting on immune cells and by modifying barrier function and microbial composition.

The temptation is to turn this into a tidy story: change the bile acids, fix metabolism. Reality is more cautious. Bile acid signaling is pleiotropic, meaning it has many effects in many tissues, sometimes pulling in different directions. FXR activation in one tissue can be beneficial while in another context it may have trade-offs. TGR5 signaling can promote energy expenditure but also influence gallbladder filling. Manipulating these pathways pharmacologically has been a lively area of research, with some promising leads and some complications. Microbial interventions add further complexity: changing the microbiome to shift bile acids might have benefits, but it could also alter colonization resistance, inflammatory tone, and lipid absorption in ways that are not uniformly desirable.

Still, the gut-liver axis is one of the clearest examples of the microbiome acting like an organ: not just a collection of organisms, but a biochemical factory integrated with host physiology. The liver sends chemical inventory into the gut; microbes remodel it; the remodeled products return as signals that influence liver function and far more besides. It is hard to think of another organ system where the input is so thoroughly host-made and the output so collaboratively edited.

A more personal way to grasp the stakes is to imagine two dinners. In one, the meal is modest in fat, rich in plant material, and eaten at a pace that lets the gut respond gradually. Bile secretion rises, does its work, and most bile acids are efficiently reabsorbed, passing through a microbial community that is not being chemically bullied. In the other dinner, the meal is very high in fat and low in fermentable plant material, repeated day after day. Bile flow becomes heavier and more frequent. Bile-tolerant microbes gain an advantage. Bile acids are modified in patterns that may amplify signaling through certain receptors and shift the hormonal landscape. Over time, this can become part of a metabolic setting in which glucose control is strained and inflammatory signals are more common. No single meal determines a fate, but patterns shape ecosystems, and ecosystems shape signals.

Even within a single person, bile acid signaling is not constant. It pulses with meals and changes with circadian rhythms. It responds to fasting and feeding, to exercise, to stress hormones, to medications. Metformin, for example, can alter bile acid handling and gut hormone secretion, and it also affects the microbiome. Statins, by lowering cholesterol synthesis, may influence bile acid synthesis indirectly. Antibiotics, as already mentioned, can shift bile transformations dramatically. The bile refinery is sensitive to daily life.

That sensitivity is also what makes bile acids attractive as biomarkers and therapeutic targets. Measuring bile acids in blood or stool can provide clues about liver function, intestinal absorption, microbial transformation capacity, and metabolic state. In principle, bile acid profiles could help identify subtypes of disease: which patients might respond to particular dietary changes, which might benefit from drugs that tweak FXR or bile acid transporters, and which might be vulnerable to infections like *C. difficile* after antibiotics. The future here is not a single magic probiotic but a more nuanced pharmacology and ecology, where bile acids are read as a functional signature of the gut-liver partnership.

There is also a quieter implication that touches everyday experience. People often treat digestion as plumbing: food goes in, nutrients are extracted, waste comes out. Bile acids make that view too small. They are part of a chemical conversation that begins with the liver, is edited by microbes, and is heard by tissues across the body. The liver is not merely reacting to what we eat; it is broadcasting molecules that will become messages. The microbes are not merely surviving; they are rewriting the broadcast. And the body is not merely tolerating it; it has built receptors that listen.

A detergent that becomes a hormone is a reminder that the gut is not a passive container. It is a chemical economy, with currencies minted by the host and exchanged, modified, and sometimes weaponized by microbes. When the exchange runs smoothly, fats are absorbed efficiently, microbial populations are kept in check, and signals help coordinate metabolism with the reality of what has been eaten. When the exchange is disrupted—by drugs, illness, diet, or chance—the consequences can appear far from the gut: in glucose curves after meals, in inflammatory markers, in susceptibility to infection, in the slow drift toward metabolic disease.

Bile acids, once imagined as the greasy assistants of digestion, turn out to be among the body's most widely traveled words. Their meanings shift depending on microbial punctuation, and the tissues that listen respond accordingly, adjusting not only how we digest a meal but how we manage energy, inflammation, and the everyday balance between comfort and distress.

4.3 The Tryptophan Fork

Some nutrients are quiet. They arrive, are absorbed, and disappear into metabolism without much drama. Tryptophan is not one of them. It is an essential amino acid, meaning the human body cannot make it and must obtain it from food. On paper, that sounds straightforward: eat protein, obtain tryptophan, build proteins. In practice, tryptophan behaves less like a brick and more like a bargaining chip. It can be spent in several different biochemical currencies, and the balance of spending shifts with diet, microbes, inflammation, and even stress.

The reason is that tryptophan sits at a literal fork in metabolism. A single molecule from a mouthful of turkey, tofu, oats, or pumpkin seeds can end up as serotonin in the gut, or be diverted into the kynurenine pathway used by the host immune system, or be transformed by microbes into indoles—small aromatic compounds that act as signals. Each route produces a different set of effects. Each route competes for the same starting material. And because the gut is where most tryptophan first meets the body, the fork is crowded: human enzymes, microbial enzymes, and immune signals all reach for the same substrate.

It helps to begin with the simplest idea: tryptophan is rare and precious. Relative to some amino acids, it is less abundant in many foods, and it is used for more than protein synthesis. The body treats it with care. But microbes treat it with curiosity. Many gut bacteria have enzymes that can catabolize tryptophan, and they do so for their own reasons: energy, redox balance, growth advantage, and chemical communication with neighbors. The host, meanwhile, uses tryptophan-derived molecules as messengers, not just within the gut but throughout the body.

The best-known tryptophan-derived messenger is serotonin, also called 5-hydroxytryptamine or 5-HT. Serotonin is popularly associated with mood, because in the brain it helps regulate emotion, sleep, and appetite. But most serotonin in the body is not in the brain. It is made in the gut, primarily by specialized epithelial cells called enterochromaffin cells. From there, it affects gut motility—the coordinated muscular contractions that move contents along—and it influences secretion, sensation, and communication with the nervous system. Some of it enters the bloodstream and is taken up by platelets, influencing clotting and vascular functions.

This is already a striking arrangement: a molecule people think of as a brain chemical is, in fact, largely a gut product. Yet the story becomes stranger: gut microbes can influence how much serotonin enterochromaffin cells make. In experiments that compared germ-free animals (raised without microbes) to colonized animals, serotonin levels in the gut were altered by microbial presence. Later work pointed to particular microbial groups—often described broadly as spore-forming bacteria—as capable of increasing host serotonin production through metabolites that stimulate enterochromaffin cells. The gut, it turns out, does not merely make serotonin; it makes serotonin in a microbial context.

If serotonin were the only outcome, tryptophan would still be interesting. But the kynurenine pathway arguably uses even more tryptophan. This is a host-driven metabolic route in which tryptophan is converted into kynurenine and downstream products, eventually feeding into the synthesis of NAD^+, a key cellular cofactor. This pathway becomes particularly prominent during inflammation. Immune signals can induce enzymes such as indoleamine 2,3-dioxygenase (IDO), which shunt tryptophan away from other uses and into kynurenine metabolism. This has immune consequences: depletion of tryptophan and accumulation of kynurenine-related metabolites can suppress certain immune responses and shape tolerance. In some contexts, this helps prevent excessive inflammation. In others, it can contribute to immune dysfunction, including the ability of tumors to dampen immune attack. The same biochemical logic—alter tryptophan availability, alter immune behavior—plays out in different arenas.

These two host pathways are important, but they are not the focus of the fork's most microbiome-specific branch: microbial indole production.

Indoles are a family of compounds built around a characteristic ring structure derived from tryptophan. Many gut microbes can convert tryptophan into indole itself or into modified indoles such as indole-3-acetic acid (IAA), indole-3-propionic acid (IPA), and indole-3-lactic acid (ILA). The names matter less than the theme: these are small molecules that can cross short distances, interact with receptors, and influence both microbial and host physiology.

One reason indoles matter is that they are intimately tied to the gut barrier. The barrier is not just a wall of cells. It is a layered defense that includes mucus, antimicrobial peptides, tight junctions between epithelial cells, and immune cells stationed beneath. A strong barrier allows the microbiome to exist in high density without continuously provoking inflammation. A weak barrier invites trouble: bacterial fragments seep across, immune cells become activated, and the gut shifts from a tolerant, regulated state to one prone to chronic inflammation.

Indoles help hold the barrier together in part through a receptor with a name that reflects its discovery rather than its everyday role: the aryl hydrocarbon receptor, abbreviated AhR. AhR was first studied because it binds environmental toxins such as dioxins, which made it seem like a sensor for industrial pollutants. Over time, it became clear that AhR is also a receptor for naturally occurring molecules, including dietary and microbial metabolites. In the gut, activation of AhR in immune and epithelial cells can promote programs that strengthen mucosal defense. One key output is the cytokine IL-22, which helps epithelial cells produce antimicrobial peptides and supports tissue repair. IL-22 is not a simple "anti-inflammatory" signal; it is better thought of as a barrier-maintenance signal, one that encourages the epithelium to stay resilient and responsive.

The historical arc of AhR is a useful cautionary tale about how biology changes meaning when placed in context. A receptor discovered through toxicology becomes, in the gut, part of ordinary physiology. The gut is a chemical environment so dense with signals that receptors are rarely dedicated to a single purpose. The same lock can accept more than one key; the same key can fit more than one lock. Microbial indoles are among the keys that help AhR do a job that looks, in retrospect, like it was always necessary: translating the presence of a thriving microbial community into signals that reinforce the boundary that makes that community tolerable.

Indoles also influence the barrier through less glamorous, more mechanical means. Some indole compounds can affect tight junction proteins, the molecular seams that seal gaps between epithelial cells. Others can influence mucus production or the behavior of goblet cells that secrete mucus. Some indoles modulate inflammation indirectly by promoting regulatory immune responses in the mucosa. The exact effects depend on the specific indole, the cell type, and the inflammatory context, but the overall message is consistent: microbial tryptophan metabolism can generate compounds that push the gut toward a well-maintained boundary rather than a leaky one.

It is worth emphasizing what makes this special compared to many other nutrients. Fiber fermentation produces SCFAs from carbohydrates the host cannot digest; bile acids are made by the host and refined by microbes; tryptophan sits at a junction where host and microbes are in direct competition for the same molecule. That competition can feel abstract until it is framed as a budgeting problem. Imagine tryptophan as a limited pool of funds. The host needs to allocate it to protein synthesis, serotonin, and NAD^+ production, while the immune system may demand more for kynurenine metabolism during inflammation. Microbes, meanwhile, take their cut to produce indoles and other metabolites. A change in one branch can squeeze the others. If inflammation induces IDO and increases kynurenine flux, less tryptophan may remain for microbial indole production and for serotonin synthesis. If the microbiome shifts toward species that aggressively catabolize tryptophan, the host may see different serotonin dynamics and different immune tone. These are not mere correlations; they are plausible consequences of a shared substrate.

A concrete example brings the competition into focus: intestinal infection and fungal overgrowth. The gut's immune system has to manage not only bacteria but also fungi such as *Candida*. Under normal conditions, fungi are kept in check by a combination of microbial competition, barrier integrity, and immune vigilance. When the barrier is compromised or immune signaling is skewed, fungi can expand and contribute to inflammation. AhR-driven pathways, including IL-22, are part of the mucosal toolkit for maintaining balanced defense. Microbial indoles that activate AhR can therefore contribute to colonization resistance—not by killing fungi directly like an antifungal drug, but by strengthening the host environment that keeps opportunists from taking advantage.

This is an important distinction. Much of modern medicine is built around direct antagonism: kill the pathogen, suppress the immune response, block the receptor. Microbiome chemistry often works through indirect ecology: change the habitat, strengthen the boundary, adjust the immune thermostat. Indoles belong to this style of control. They are not antibiotics. They are more like maintenance signals that make the gut less hospitable to invasion and less prone to inflammatory overreaction.

The microbial world, of course, uses indoles for its own purposes as well. Indole is a well-known bacterial signaling molecule. In some species it influences biofilm formation, motility, and stress responses. That means indoles are bilingual: they can serve as microbial-to-microbial messages and microbial-to-host messages. This dual role helps explain why indole metabolism is often conserved. A molecule that helps microbes coordinate with each other and also shapes the host environment in a favorable way is a powerful evolutionary tool.

A particularly intriguing wrinkle is cross-feeding among microbes. Not every bacterium produces the same indoles, and some metabolites can serve as stepping stones. For example, certain *Lactobacillus* species can produce indole-3-lactic acid (ILA). In experimental settings, ILA can support broader production of other indole derivatives through microbial networks, effectively nudging the community's chemical output. This is reminiscent of how acetate can serve as a feedstock for butyrate production: a community can behave like a metabolic assembly line. In a gut ecosystem, the final message may not belong to a single species; it may be a collective product.

This collective aspect matters for interventions. People often ask whether taking tryptophan supplements could boost helpful indoles or improve mood via serotonin. The fork makes such logic uncertain. More substrate does not guarantee more of a desired product, because enzymes and pathways can become saturated or diverted.

If inflammation is high, additional tryptophan may be preferentially routed into kynurenine metabolism. If the microbiome is depleted by antibiotics, the microbial branch may be diminished regardless of substrate availability. If certain key indole-producing genes are absent, the pathway cannot be conjured by wishful eating. Dietary change is powerful, but it does not operate like turning a single knob.

A more realistic takeaway is that the tryptophan fork is sensitive to context and can help explain why gut symptoms, immune symptoms, and even some aspects of neurological well-being can cluster. Consider a person under chronic stress who develops irregular bowel habits and abdominal discomfort. Stress hormones can alter gut motility, which changes microbial habitat and substrate availability. Motility changes can affect how long tryptophan and other nutrients remain in contact with microbes. Stress and disrupted sleep can also influence immune tone, potentially shifting tryptophan flux toward kynurenine pathways. At the same time, dietary patterns under stress often change—less fiber, more processed foods, fewer varied proteins—which can reshape the microbiome and its capacity to produce indoles and SCFAs. None of this proves a single causal chain, but it provides a credible web of interactions: altered motility, altered microbial metabolism, altered signaling molecules, altered barrier, altered immune reactivity, and altered perception of gut sensations. The gut is a place where chemistry and experience meet.

There is a second caution here. Because indoles can activate AhR, and AhR has a reputation as a receptor for toxins, it is tempting to imagine that "more AhR activation" is always good or that indoles are always protective. Biology resists such simplification. Different AhR ligands can produce different patterns of activation. Duration matters: a brief pulse after a meal is not the same as chronic stimulation. Cell type matters: what helps barrier maintenance in one context could have different consequences in another. The goal in physiology is rarely maximal activation; it is appropriate activation, timed and balanced. Microbial communities, when stable and diverse, tend to produce a repertoire of metabolites that may keep signaling within a healthy range. When communities are disrupted, signaling can become skewed.

Even with these caveats, indole production remains one of the most elegant examples of microbial metabolism serving the host. It turns an amino acid into a set of molecules that help the gut defend itself without constant warfare. It provides a mechanism by which diet and microbes can influence immune programs such as IL-22 that are central to barrier resilience. It connects the microbial ecosystem to host neuroendocrine outputs like serotonin, not by direct mind control but by acting on the gut cells that make most of the body's serotonin. And it reveals why inflammation can have cascading effects: the immune system's decision to route tryptophan into kynurenine metabolism is not merely a metabolic footnote; it is a choice that can change microbial chemistry and barrier behavior.

The fork also reframes what "protein" means in diet. Protein is not only about muscle or satiety. Its amino acids are substrates for chemical messages. A diet rich in varied plant and animal proteins, delivered alongside fermentable fibers that support a diverse microbiome, offers not just nutrition but a broader palette for microbial metabolism. The goal is not to chase a single molecule but to support a stable, communicative ecosystem whose metabolites reinforce the barrier and keep immune responses proportionate to the threats at hand.

When tryptophan is plentiful enough for the host to meet its needs and for microbes to produce indoles, the gut tends to behave as though it expects coexistence: epithelial cells repair quickly, antimicrobial peptides are produced where needed, immune cells patrol without panicking, and motility keeps the ecosystem from stagnating. Under those conditions, the chemical fate of a single amino acid becomes part of a larger, reassuring pattern: microbes take what we cannot use directly, refine what we can, and return metabolites that help the body maintain a boundary strong enough to allow life at high microbial density.

Chapter 5

The Peacekeepers

Far from being a simple killing machine, the immune system acts as a sophisticated diplomatic corps trained by the very microbes it monitors. This chapter reveals how our resident bacteria educate immune cells to distinguish friend from foe, preventing the body from attacking itself. We uncover the mechanisms of 'colonization resistance'-where microbes physically block invaders-and the catastrophic ecosystem collapse that occurs when antibiotics strip this invisible armor away.

5.1 The School of Tolerance

A newborn arrives with an immune system that is powerful but inexperienced. It can build armies of cells, manufacture molecular weapons, and remember enemies for decades. Yet it begins life with a central problem: most of what it will meet is not an enemy. Food proteins, pollen dust, skin flakes, the bacteria that take up residence in the gut and on the skin—these are constant companions, not invaders. If the immune system treated every unfamiliar molecule as a threat, the body would live in a perpetual state of alarm. Allergies and autoimmune diseases can be read, in part, as failures of restraint: the alarm system is too sensitive, or it never learned which noises are ordinary.

The microbiome helps teach that restraint. Not by giving the immune system a list of safe substances, like a school handbook, but by shaping the conditions under which immune decisions are made. Many of the most consequential immune choices are made at borders: the gut lining, the airways, the skin. These places are not sterile hallways; they are busy marketplaces where molecules and microbes press close to the body's tissues. The immune system's challenge is not simply to identify what is "foreign" but to decide what kind of response is warranted. That decision is less like a switch and more like a dial: the same antigen can provoke tolerance in one setting and inflammation in another.

A useful place to start is with T cells. These are white blood cells that do much of the immune system's thinking. Each T cell carries a receptor that can recognize particular molecular shapes. The body generates a huge variety of these receptors at random. This randomness is the point: it ensures that, somewhere in the crowd, there will be a T cell capable of recognizing almost any pathogen the world might throw at us. The price is that many T cells will also be capable of recognizing harmless targets—including, occasionally, the body's own molecules.

The thymus, an organ above the heart, does the first round of education. It eliminates many dangerous T cells that react strongly to self. But the thymus cannot anticipate the whole menu of harmless things a person will encounter across a lifetime, and it cannot perfectly remove every self-reactive cell without also crippling immune diversity. So the immune system relies on a second kind of schooling that happens out in the body, especially at barrier surfaces. This is where the microbiome becomes a tutor.

The most important graduates of this second school are called Regulatory T cells, often shortened to Tregs. "Regulatory" here means peacekeeping: these T cells dampen immune responses, prevent runaway inflammation, and help the body coexist with harmless antigens. A key marker of many Tregs is a gene program controlled by a transcription factor called Foxp3. When naïve T cells (young, uncommitted cells) receive certain signals, they can become Foxp3-positive Tregs. Under different signals, the same naïve cells can become other types of helper T cells designed for attack.

Among the attack-oriented lineages, one has become especially central to thinking about the microbiome: Th17 cells (T helper 17), named for a signaling protein they produce called interleukin-17. Th17 cells are useful at mucosal surfaces; they help recruit defenses against particular bacteria and fungi and help strengthen barrier responses. But they are also associated with inflammatory diseases when the balance tips too far toward aggression. In many ways, Th17 cells and Tregs form a conceptual yin and yang: protection can mean fighting, or it can mean calming the fight.

How does a naïve T cell decide which path to take? It does not decide alone. It is coached by dendritic cells, professional antigen-presenting cells that function like intelligence officers. Dendritic cells sit beneath epithelial barriers and continually sample what is happening above them. In the gut, they can extend slender projections between epithelial cells to capture bits of bacteria or food. They can also receive samples that are ferried across the epithelium by specialized cells and by transport mechanisms that move small amounts of material from the lumen into the immune tissue below. Sampling is normal. The gut is not a sealed fortress; it is a negotiated border.

When dendritic cells capture antigens, they process them and display fragments on molecular billboards called MHC (major histocompatibility complex). Naïve T cells scan these billboards. Recognition is only the first step. The dendritic cell also provides contextual cues: co-stimulatory signals and cytokines (small messenger proteins) that tell the T cell whether this encounter should be treated as a crisis or as an ordinary introduction.

Those contextual cues are influenced by microbes. Some microbes, or their molecules, encourage dendritic cells to adopt a calming style, one that nudges naïve T cells toward becoming Tregs. Other microbes, or the inflammatory signals that accompany their invasion, push dendritic cells toward a more alarmist stance that favors Th17 differentiation. The outcome is not just "microbe A creates Tregs" and "microbe B creates Th17." The same microbial products can have different effects depending on the surrounding chemical weather: the presence of tissue damage, the mix of cytokines, the nutritional environment. Context is everything, which is why the immune system can tolerate trillions of bacteria on an ordinary day and still mount a fierce defense when those bacteria cross boundaries or when a true pathogen arrives.

A historical shift in thinking helped make this picture possible. For much of the twentieth century, microbes were chiefly framed as enemies. This was understandable: antibiotics transformed medicine, and infectious diseases had an obvious villain. But as allergic and autoimmune diseases rose in many industrialized countries, researchers began to suspect that a too-clean environment might carry its own risks. One influential framing, associated with David Strachan's 1989 observation that children from larger families had lower rates of hay fever, proposed that early-life infections might protect against allergies. Over time, this idea was refined into what became known as the "old friends" hypothesis, associated particularly with Graham Rook and others in the early 2000s: the immune system evolved alongside a set of microbes and parasites that were once common companions, and losing exposure to this ecological context could leave immune regulation undertrained. The point was not that children needed measles or whooping cough. It was that immune tolerance might require a steady stream of microbial signals that say, in effect, "Most encounters are not emergencies."

The gut microbiome provides precisely such a stream. Babies begin acquiring microbes at birth and in the early months of life, through contact with caregivers, breastfeeding, and the environment. During this period, the immune system is not passive. It is actively calibrating the inflammatory dial. The gut-associated lymphoid tissue—immune structures embedded in and around the intestines—is one of the largest immune organs in the body, and it is positioned where diet and microbes constantly deliver lessons.

One of the most vivid lessons comes from microbial metabolites: small molecules produced when microbes digest components of our diet. Among the most important are short-chain fatty acids (SCFAs), including acetate, propionate, and butyrate. These molecules are especially tied to the fermentation of dietary fiber—carbohydrates that human enzymes cannot fully break down. When gut bacteria ferment fiber, they produce SCFAs that bathe the colon.

Butyrate has become a kind of celebrity in microbiome immunology because it connects diet, microbes, and immune tolerance with unusual clarity. In 2013, Furusawa and colleagues showed that butyrate can promote the differentiation of colonic Foxp3-positive Tregs in mice and can ease inflammation in a model of colitis. The key idea is not merely that butyrate "reduces inflammation" but that it helps steer T cells toward a regulatory identity. Mechanistically, butyrate can act as an inhibitor of histone deacetylases (HDACs), enzymes that influence how tightly DNA is packaged and which genes are accessible for transcription. When HDAC activity is inhibited in certain contexts, genes associated with regulatory programs can become more readily expressed. In the language of everyday life: a microbial metabolite can change how easy it is for a T cell to read the pages that contain peacekeeping instructions.

This does not mean SCFAs are simple sedatives. In 2015, Park and colleagues emphasized context-dependence: SCFAs can promote effector responses such as Th1 or Th17 under some inflammatory conditions, while still supporting regulatory programs in others. This apparent contradiction is exactly what a well-calibrated immune system needs. A border guard who always smiles is useless; one who always shoots is disastrous. The immune system uses microbial metabolites as inputs, not as orders. The same molecule can contribute to tolerance at baseline and to effective defense during infection, depending on the surrounding signals.

Another example shows that microbes do not only teach through broad chemical atmospheres; sometimes they use specific molecular phrases. In 2008, Sarkis Mazmanian's group reported that a molecule from *Bacteroides fragilis*, a common gut commensal, could protect mice from experimental colitis. The molecule was polysaccharide A (often abbreviated PSA), a complex sugar on the bacterial surface. PSA helped drive anti-inflammatory immune pathways, including interleukin-10 (IL-10), a cytokine strongly associated with immune restraint. Later, Monteleone and colleagues (2015) extended related ideas to human cells in culture, showing that PSA could promote regulatory features in human T cells and that dendritic cells were required for the effect. The striking implication is that some commensals are not merely tolerated because they hide well; they participate in maintaining tolerance by speaking directly to the immune system.

At this point, it is tempting to imagine a simple morality play: friendly bacteria induce Tregs; hostile bacteria induce Th17. Nature is messier. A dramatic demonstration comes from segmented filamentous bacteria (SFB), unusual microbes that adhere closely to the intestinal epithelium in some animals. In 2009, Ivanov and colleagues showed that colonization with SFB could induce Th17 cells in the small intestine of mice. In that model, the Th17 response improved resistance to a gut pathogen, *Citrobacter rodentium*. In 2014, Goto and colleagues linked this effect to antigen presentation by intestinal dendritic cells: SFB antigens presented via MHC class II were crucial for Th17 induction. Here, a commensal-like resident pushes the immune system toward a state that looks, on paper, more inflammatory. Yet the result can be protective. The boundary between "commensal" and "pathogen" is not always a fixed label. Sometimes the immune system chooses a heightened defensive stance not because something is evil, but because the ecological relationship requires careful control at a particular interface.

This brings us back to tolerance. Tolerance is not ignorance. It is an active program of controlled exposure and measured response. Nowhere is that clearer than in the process called oral tolerance, the tendency of the immune system to become unreactive to antigens encountered through the gut. When this process works, food proteins do not spark chronic inflammation; they become part of the background of daily life. When it fails, the consequences can be immediate and frightening, as in food allergies that trigger anaphylaxis.

The microbiome influences oral tolerance in several ways. It affects how the gut barrier functions: the thickness and properties of mucus, the integrity of tight junctions between epithelial cells, the production of antimicrobial peptides. It affects how antigens are processed and presented, and how inflamed the tissue environment is when presentation occurs. It also affects the development and stability of Tregs themselves. Some Tregs are generated in the thymus, but many are induced in peripheral tissues like the gut, where the immune system constantly negotiates with outside material. The gut's Tregs are not a generic peacekeeping force; they can be specialized to the gut's unique demands, shaped by signals that include microbial products.

Thinking of this as a school helps because schools are not only about lessons; they are about timing and repetition. Early life appears to be a particularly sensitive period. The immune system is developing its baseline settings. The microbial community is assembling. Diet is changing rapidly. The two systems are learning each other's rhythms. Disturbances in this period—through severe infections, altered diets, or courses of antibiotics—can have outsized effects. Not because a single perturbation writes an irreversible fate, but because the early months and years are when the immune system is most actively deciding what to treat as normal.

To make this vivid, imagine a naïve T cell as a new employee at a security agency. It has a weapon and a badge but no experience. A dendritic cell is the supervisor who hands it a folder: a piece of antigen. Along with the folder comes a tone of voice and a workplace culture. Is everyone shouting, sirens blaring? Or is this a routine briefing? Microbes shape that culture by shaping the chemical environment: SCFAs, bile acid derivatives, vitamins, and a host of other metabolites; they also shape the pattern of receptor stimulation that dendritic cells experience through their innate sensing machinery. Over time, the security agency develops habits. It learns which situations call for a calm report and which call for a raid.

The key decision point—Treg versus Th17, tolerance versus inflammatory readiness—is not made on moral grounds. It is made on evidence. The immune system uses proxies: barrier integrity, signals of tissue damage, patterns associated with invasion, and the presence of regulatory metabolites. One reason allergies and autoimmune diseases are so difficult is that these proxies can be misleading. If the barrier is chronically irritated, the immune system may treat harmless antigens as if they arrived with danger. If the regulatory signals are weak, the immune system may fail to stabilize tolerance even when the antigen is innocuous.

A helpful everyday example is the difference between a calm gut and an inflamed one. During a stomach bug, the lining is disrupted, cells are stressed, and inflammatory cytokines are abundant. Food proteins encountered in that setting can be presented to the immune system in a context that looks dangerous. Under ordinary conditions, the same proteins are presented amid signals that favor tolerance. The microbiome contributes to those ordinary conditions by helping maintain barrier integrity and by providing metabolites that support regulatory pathways. This is one way to understand why the timing of exposures matters in early life: encountering antigens during a period of stable, diverse microbial signals may train tolerance, while encountering them during repeated inflammatory disruptions may do the opposite.

The immune system's education is also a matter of geography. The gut is layered: the lumen where microbes reside, the epithelial barrier, and the immune-rich tissue beneath. Dendritic cells operate at the boundary between the visible world of microbes and the internal world of immune decision-making. They are also translators: they convert complex microbial and dietary information into a small set of instructions that T cells can act on. Because this translation step is so central, it is a place where the microbiome can exert enormous influence without ever "invading" the body.

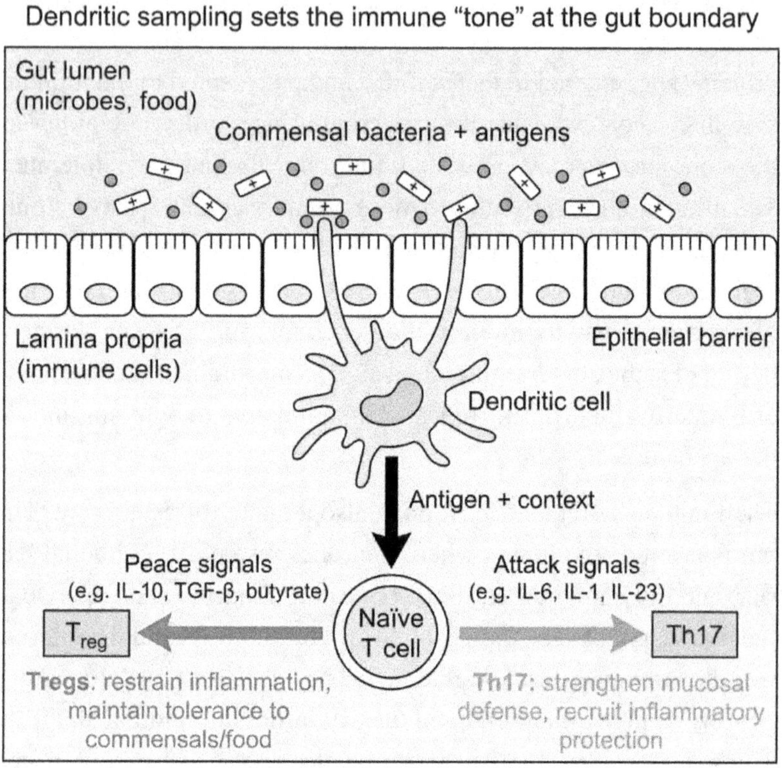

The diagram captures a decision that is made countless times a day. A dendritic cell samples the gut lumen, processes what it finds, and presents it to a naïve T cell. The same antigen can lead to different outcomes. When the surrounding signals suggest stability and partnership—often reinforced by microbial metabolites and regulatory cytokines—the T cell is nudged toward a regulatory fate, becoming part of the peacekeeping force that prevents unnecessary inflammation. When the signals suggest danger—often marked by inflammatory cytokines and patterns associated with invasive behavior—the T cell is nudged toward an effector fate such as Th17, strengthening defenses at the surface.

What makes early life special is that these decisions are not happening against a fixed backdrop. The cast of microbes is changing; the immune system is expanding its repertoire; tissues are developing their baseline barrier properties. Small differences in the early microbial environment can, in principle, shift the probabilities of these fate decisions, and repeated small shifts can accumulate into a different immune temperament. The idea is not determinism but calibration. A thermostat set slightly too high may never cause trouble on a mild day, but it will matter during a heat wave.

This calibration reframes tolerance as something earned through daily practice. The immune system must repeatedly encounter harmless antigens in contexts that teach calm. It must also learn the opposite lesson: that true pathogens and tissue injury require forceful responses. The microbiome helps provide both lessons, not as a script but as a living environment that supplies signals, shapes barriers, and keeps the immune system's training realistic. The result, when it goes well, is not an immune system that is sleepy or naïve. It is an immune system that can afford to be relaxed most of the time because it is not confused about what counts as danger.

5.2 The Invisible Shield

A healthy gut does not feel like a battlefield. Most days it feels like nothing at all: a quiet conveyor belt where breakfast becomes blood sugar, building blocks, and waste. That silence is an achievement. The intestine is a warm tube lined with nutrients, constantly seeded by microbes from food and from other people, connected to a world in which diarrheal pathogens have always been a fact of life. If the body relied only on emergency responses—inflammation, fever, antibodies arriving after an alarm is raised—it would be forever late. Much of the protection happens earlier and more subtly, through a kind of prevention that looks less like combat than like occupancy.

Microbiologists sometimes call this *colonization resistance*. The phrase is dry, but the idea is intuitive: established residents make it hard for newcomers to settle. In the gut, the residents are mostly commensal bacteria, members of a community that has already divided up space, claimed food sources, and laid down the chemical rules of the neighborhood. The result is an invisible shield: protection created by a crowded ecosystem.

What makes this hard to picture is that "crowding" in the gut is not like crowding in a room. The intestine is layered. There is the lumen, the central channel where food is digested and where most microbes live. There is the mucus layer coating the intestinal wall, a slippery hydrogel made by the host that traps particles, slows diffusion, and provides a habitat of its own. Then there is the epithelial sheet, a single-cell-thick lining that must absorb nutrients without becoming a welcome mat for invaders. Beneath that sheet lies immune tissue, ready to respond if the border is breached. The invisible shield works across these layers, and it works by using the simplest ecological strategies: taking up space, eating the available food, and changing the environment so that interlopers find it inhospitable.

Consider the gut as a city where every niche is a job. Some bacteria specialize in breaking down complex carbohydrates; others consume the breakdown products; others use oxygen traces near the epithelium; others are adapted to the mucus layer. In a stable community, those jobs are filled. A pathogen arriving from the outside has to do more than survive passage through stomach acid. It must also compete. It must find a nutrient it can use better than the residents. It must attach to a surface that is not already occupied or defended. It must resist the local chemical climate. When the community is intact, these are tall orders.

A concrete example comes from one of the most feared opportunists in modern hospitals: *Clostridioides difficile.* It is not a classic invader that needs to be caught from a sick person in the usual way. Many people encounter its spores, and spores can persist in the environment. The real vulnerability often appears after antibiotics. A person takes a broad-spectrum antibiotic for a legitimate reason—pneumonia, a surgical infection, a dental abscess—and a few days later develops severe diarrhea and colitis. The intuitive assumption is that antibiotics should help, because they kill bacteria. The paradox is that they also kill the bacteria that were preventing *C. difficile* from gaining traction.

The details of that protection reveal how ecological the invisible shield really is. The colon is not just full of microbes; it is full of microbial *functions*. One function involves bile acids. The liver produces bile acids and secretes them into the intestine to help digest fats. Many of these are "primary" bile acids. Gut microbes chemically modify them into "secondary" bile acids. This matters because *C. difficile* uses bile acids as signals. In experimental systems, certain primary bile acids can promote spore germination, while certain secondary bile acids can inhibit growth and outgrowth. After antibiotics, the microbial community that performs these conversions is disrupted. The chemical landscape shifts: the balance tilts toward primary bile acids and away from inhibitory secondary ones. In work led by Eric Pamer's group (Buffie and colleagues, 2015), a specific commensal bacterium, *Clostridium scindens*, was associated with resistance and could enhance resistance in model systems in a secondary-bile-acid-dependent manner. Protection, in other words, was not just about numbers of bacteria; it was about a metabolic service performed by particular members of the community.

This is an example of a broader principle: colonization resistance is not simply the microbiome "fighting" pathogens. It is the microbiome maintaining an environment in which pathogens struggle to execute their life cycle. The residents do not need to recognize *C. difficile* as an enemy. They only need to keep doing what they were doing—occupying niches, consuming substrates, producing metabolites that make the gut feel like home to them and hostile to outsiders.

The most direct mechanism is nutrient competition. Pathogens are not ghosts; they are living organisms with appetites. Many enteric pathogens bloom when they find a particular nutrient suddenly abundant. Antibiotics can create that abundance by removing competitors. Food arrives in the gut partly digested, and microbes finish the job. When a large portion of the community is wiped out, there are unclaimed calories. Some of these are simple sugars; others are amino acids; others are compounds released from mucus or from dead microbes. The details vary by pathogen, but the logic does not: empty pantries rarely exist in a stable gut, and antibiotics can stock them.

Competition is not only about what is eaten, but where. Many pathogens need to get close to the epithelium to cause disease. They may attach to receptors on epithelial cells, inject toxins, or exploit small breaches. But the epithelium is not bare. Commensal microbes are often physically positioned in ways that block access, especially in the outer mucus layer. Some commensals form dense, mucus-associated communities that behave like biofilms—structured microbial neighborhoods embedded in a matrix. These communities can exclude certain groups of bacteria simply by being there first and by maintaining local conditions that favor them. Even when the gut lumen contains a diverse soup of microbes, the mucus layer can be a distinct habitat with its own rules, and its integrity can depend on particular commensals. A pathogen that thrives in the free-flowing lumen may find the mucus zone difficult to penetrate if the resident community is intact.

The host contributes to this physical organization. The intestine secretes antimicrobial peptides, mucus, and immunoglobulin A (IgA), an antibody that is released into the gut lumen and can coat microbes, affecting their movement and attachment. There are also host proteins that act as ecological regulators, shaping where bacteria can live without necessarily killing them outright. One example is RegIIIγ, an antimicrobial lectin that in the small intestine helps maintain a bacteria-free zone close to the epithelial surface—a gap on the order of tens of micrometers. This spatial separation matters. When bacteria are kept at a respectful distance, the epithelium can absorb nutrients and sense the world without being constantly provoked. When the separation collapses, even harmless residents can trigger immune activation simply because they are too close. The invisible shield is therefore partly a *spacing policy*: a managed buffer zone that allows coexistence without constant confrontation.

The microbiome, in turn, reinforces this spacing. Many commensals stimulate mucus production and help maintain epithelial health by producing metabolites that feed epithelial cells. Butyrate, introduced earlier as a signal that can support regulatory immune programs, is also a preferred fuel for colonocytes (colon epithelial cells). When colonocytes have sufficient butyrate, they can maintain tight junctions and barrier functions more effectively. A well-fed epithelium is less leaky. That alone makes invasion harder. The shield, again, is indirect: it is not a row of soldiers but an infrastructure project.

Chemical warfare does exist, but it is usually not the dramatic kind imagined in popular culture. Many bacteria produce bacteriocins: proteins or peptides that can kill or inhibit closely related bacterial strains. These are often narrow-spectrum weapons, more like targeted assassinations than carpet bombing. They can allow resident strains to defend their niche against competitors. From the host's point of view, this microbial infighting can be useful: it keeps the community stable and can suppress potential pathogens that are close relatives of residents. But it is also a reminder that the microbiome is not a unified army working for us. It is a complex society in which our health can be a side effect of microbial self-interest.

If colonization resistance is so powerful, why do pathogens ever succeed? One reason is that pathogens are opportunists that exploit disturbance. Antibiotics are a common disturbance in modern life, but they are not the only one. Sudden dietary shifts can alter the availability of key substrates. Gastrointestinal infections can damage the mucosa and change oxygen levels near the epithelium. Inflammation itself can reshape the habitat. During inflammation, the host releases reactive molecules and changes nutrient flows, and this can inadvertently favor certain bacteria, including some that are associated with disease. A pathogen that can thrive under inflammatory conditions may, perversely, benefit from provoking inflammation. In that sense, some pathogens do not merely invade an ecosystem; they *engineer* it.

A historical anecdote illustrates how medicine came to appreciate the protective role of resident microbes. For decades, clinicians noticed that antibiotic treatment sometimes led to severe colitis, originally called pseudomembranous colitis because of the characteristic plaques seen in the colon. The association with antibiotics was clear, but the culprit organism was not always. In the late 1970s, *C. difficile* was identified as a major cause, and the idea that an antibiotic could cause a dangerous infection by disrupting the normal flora gained force. This was a conceptual pivot. Antibiotics did not simply remove enemies; they could dismantle a protective ecosystem. The gut was not merely a place where bacteria *protected.*

Colonization resistance also helps explain why many "bad" bacteria do not inevitably cause disease. Take *Salmonella.* People can ingest it and never develop severe illness, or develop a mild case, while others become dangerously ill. Dose matters. Host factors matter. But so does the background ecosystem. A diverse community can prevent *Salmonella* from blooming by limiting the nutrients it needs and by maintaining barrier functions that reduce its access to tissue. When the ecosystem is disrupted, the same exposure can lead to a different outcome.

This ecological framing changes how one thinks about hygiene and modern living. The goal is not to be dirty. The goal is to avoid creating a body landscape that is perpetually disturbed. The old friends idea, in its most practical form, is not a call to romanticize parasites or pre-industrial life. It is a reminder that immune health depends on stable relationships with microbes and on the maintenance of the ecological services they provide. Overuse of antibiotics, overly restrictive diets low in fermentable fibers, and other forces that reduce microbial diversity or function can erode those services. In the gut, erosion can look like a vacancy sign: niches unfilled, nutrients unclaimed, chemical barriers weakened.

The invisible shield is also a lesson in humility for anyone tempted by simple interventions. If protection is an emergent property of a community—a composite of who is present, what they are doing, and where they are located—then adding a single microbial strain as a probiotic cannot be expected to replicate the effect, at least not reliably. Sometimes a single function does matter, as with bile acid metabolism and *C. difficile*. But often the protective effect depends on networks: cross-feeding relationships where one microbe's waste is another's food; spatial arrangements where mucus-associated strains create a habitat that excludes opportunists; host responses that maintain separation zones and mucus thickness. An ecosystem cannot be reduced to a mascot.

At the same time, the idea is not mystical. It is practical ecology. A crowded, well-functioning microbial community makes the gut harder to invade for the same reason a well-run city is hard to take over quietly: resources are accounted for, spaces are occupied, and the environment is shaped by established routines. The gut's residents do not stand guard in uniforms. They simply live there so successfully that, for many would-be invaders, there is nowhere to stand and nothing easy to eat.

5.3 Scorched Earth

The story of antibiotics is one of modern medicine's proudest legends, and for good reason. Before these drugs, a cut finger could become a death sentence, childbirth was haunted by sepsis, and pneumonia routinely carried people off in a matter of days. When penicillin began saving soldiers and civilians in the 1940s, it did not merely add a new treatment; it changed what it meant to be vulnerable. Many of the most frightening microbes were demoted from inevitabilities to inconveniences.

That success created an understandable habit of mind: bacteria are the enemy, antibiotics are the weapons. Yet the body is not a sterile castle occasionally attacked by germs. It is an ecosystem in which microbes are permanent residents, and some of the protection described earlier depends on their steady presence. When antibiotics enter this world, they behave less like a sniper and more like a wildfire. They are designed to damage bacterial cells, not to negotiate with the microbial community about which members are helpful. Even the most carefully prescribed course is, from the microbiome's perspective, a sudden ecological catastrophe.

The phrase "broad-spectrum" sounds reassuring, like a wider net cast to catch the culprit. Ecologically it means something else: collateral damage. The gut microbiome contains hundreds of species, many of them hard to culture and poorly known by name, but collectively performing a vast set of functions. They ferment fiber into short-chain fatty acids, transform bile acids, compete for nutrients, build spatial structure in mucus, and help tune immune tone at the border. Antibiotics can strip away not only particular species but whole functional guilds. What remains after treatment is rarely the same community, and the path back to stability can be uneven, incomplete, and deeply personal.

A familiar human scene makes the problem concrete. Someone develops a urinary tract infection and receives an antibiotic. The infection clears; the fever goes away; life resumes. Then, a week later, they develop watery diarrhea, cramping, and a distinctive, sickly smell that clinicians learn to recognize. A stool test comes back positive for *Clostridioides difficile*. The patient is bewildered: they took medicine to kill bacteria, and now they have an infection in the gut. This is not a rare plot twist. It is a signature of microbiome disruption.

C. difficile is a master opportunist because of its spores. Spores are tough, dormant forms that can survive cleaning agents, dryness, and time. They can sit quietly in the gut or in the environment until conditions become favorable. Antibiotics help create those conditions in several complementary ways.

One way is brutally simple: fewer competitors. Many gut bacteria that normally occupy niches and consume nutrients are sensitive to common antibiotics. When they are removed, unclaimed food sources appear. The newly vacant ecosystem becomes a buffet. *C. difficile* can then expand rapidly, producing toxins that damage the lining of the colon and provoke inflammation. This damage can further reshape the habitat in ways that favor the invader, creating a vicious loop of injury and bloom.

Another way is chemical, and it shows how antibiotics can flip the gut from "hostile" to "welcoming" without changing anything about the patient's diet. Bile acids, produced by the liver and secreted into the gut, are not just detergents for fat digestion; they are ecological signals. The microbiome normally modifies bile acids into secondary forms. Experimental work by Jennifer Theriot and colleagues (2014) and subsequent studies emphasized that antibiotics can shift both the microbial community and the metabolome, including bile acid profiles. In model systems, increases in primary bile acids such as taurocholate can promote *C. difficile* spore germination, while the depletion of certain secondary bile acids removes inhibitory pressure on outgrowth. There is a kind of cruel elegance here: antibiotics do not merely open space; they change the signage. They replace "Do not enter" signals with "Welcome" signs.

The geography of the gut makes the situation even more interesting. Evidence from experimental systems suggests that spore germination can be supported in small intestinal contents regardless of antibiotic exposure, while large-intestinal outgrowth is more tightly controlled by microbiota-dependent factors such as secondary bile acids. In other words, some steps in the *C. difficile* life cycle may be easy to start, but hard to finish when the microbial community is intact. Antibiotics weaken the controls in the large intestine, turning a contained event into a full-blown occupation.

The collateral damage extends beyond *C. difficile*. Antibiotics can enable domination by other opportunists, including antibiotic-resistant organisms. A particularly sobering example comes from vancomycin-resistant *Enterococcus* (VRE), a group of bacteria that can live quietly in the gut and then cause dangerous bloodstream infections in vulnerable patients. In mouse experiments, antibiotic treatment can allow VRE to dominate the intestinal community. In clinical observations, intestinal domination has been linked to subsequent bloodstream infection: the gut becomes not just a reservoir, but a launchpad. The threat is not only that resistant bacteria exist, but that ecological disruption gives them the stage.

How long does the microbiome take to recover after antibiotics? The honest answer is: it depends, and it varies widely. In humans, the post-antibiotic microbiome often does not simply snap back to a previous state, especially after repeated courses. Some people recover much of their former diversity within months; others show lasting shifts. The reasons include differences in diet, genetics, exposures, the starting community, and sheer chance. The gut is reseeded from food, environment, and close contacts, but it is also shaped by the habitat that remains. After a fire, the species that return first can change which forest grows back. Ecologists call this a *priority effect*. The microbiome has its own versions of that phenomenon. Early recolonizers can claim niches and alter the environment, making it easier for some microbes and harder for others. A single antibiotic course can therefore push a person into a new stable configuration that is not obviously "sick" but may be less resilient to future disturbance.

Resilience is the key word. A resilient ecosystem can absorb a shock and return to function. Antibiotics can reduce resilience by narrowing diversity and removing functional redundancy. Functional redundancy means that multiple species can perform similar tasks. If one is lost, another can cover the job. When redundancy is high, the system is robust. When redundancy is low, the loss of a single group can create a functional gap, like losing all the electricians in a city at once. For colonization resistance, those gaps might involve bile acid conversion, fiber fermentation, or the ability to occupy a particular spatial niche in the mucus layer.

This is where the metaphor of scorched earth earns its place. A scorched-earth strategy in war denies resources to an enemy by destroying the land itself. Antibiotics, by wiping out large swaths of the microbiome, can deny pathogens competitors, but at the cost of damaging the protective landscape. The intended target may be a bacterium in the lungs or the skin, yet the gut pays a heavy price simply because it contains so many bacteria and because many antibiotics reach it directly or indirectly. Even drugs taken for non-gut infections can end up bathing the intestinal community through biliary secretion or incomplete absorption.

The historical irony is that antibiotics were initially celebrated not only for curing infections but also for their apparent safety and simplicity. Early in the antibiotic era, there were already hints that the normal flora mattered, but the tools to measure it were crude. Culture-based microbiology could count certain bacteria, but most gut microbes do not grow easily on standard plates. The microbiome remained largely invisible. It is hard to worry about what you cannot see. Only with modern DNA-based methods did the extent of antibiotic-induced disruption become quantifiable, and with metabolomics did the functional consequences become measurable. Suddenly, the gut after antibiotics could be seen not as "cleaner" but as ecologically transformed.

The emotional stakes are clearest in patients who experience recurrent *C. difficile* infection. The first episode can be treated, but recurrences are common, and each recurrence can further destabilize the microbiome. Patients describe a peculiar dread: not only of the symptoms, but of the sense that their own gut has become untrustworthy, a place where a microscopic opportunist can erupt without warning. The illness is not merely diarrhea; it is loss of bodily confidence, a shrinking of life around proximity to a bathroom, a fear of the next prescription.

Medical practice has adapted. Clinicians now think carefully about antibiotic stewardship: using antibiotics only when needed, choosing the narrowest effective agent, and limiting duration. This is not just a public health strategy to slow resistance; it is a personal ecological strategy to preserve the patient's microbial defenses. The goal is not to withhold life-saving drugs. It is to treat them with the respect one gives to a powerful tool that can also cause harm.

The emerging research frontier is rebuilding after damage, not by guessing, but by restoring missing functions. One can see the shift from generic "take a probiotic" advice toward mechanistic interventions. If susceptibility to *C. difficile* is partly driven by a loss of bile-acid-transforming capacity, then a rational strategy is to restore that capacity. Work associated with Pamer's group (Buffie and colleagues, 2015) exemplified this idea by pointing to *Clostridium scindens* and bile acid metabolism as a protective function that could be reconstituted in model systems. The larger promise is not that a single bacterium will solve every post-antibiotic problem, but that understanding *what was lost* allows a more disciplined approach to repair.

Repair is complicated by the fact that antibiotics do not only remove microbes; they can alter the host environment itself. Inflammation, changes in mucus, altered gut motility, and shifts in immune signaling can persist after the drug is gone. A recovering microbiome returns to a habitat that may not be identical to the one it left. That helps explain why repeated antibiotic courses can have cumulative effects. Each perturbation changes the baseline from which the next begins.

None of this negates the miracle of antibiotics. It sharpens it. A miracle that saves lives on one front can quietly weaken defenses on another. The body can tolerate that trade-off remarkably well most of the time, which is why antibiotics remain among the most valuable medicines ever developed. But when the trade-off goes wrong—when the microbial peacekeepers are removed and an opportunist takes advantage—the consequences reveal how much protection was being provided by an ecosystem that most people never knew they had.

A person who has never had *C. difficile* may imagine the gut microbiome as a passive background, like wallpaper. After a severe post-antibiotic infection, it is easier to see it as architecture: load-bearing structures, invisible supports, and critical pathways. Antibiotics can knock holes in that structure. The body will often repair them, but the repair is not always complete, and the period before stability returns can be a dangerous opening. In that opening, microbes that were once held in check do not need to become more aggressive. They only need to do what opportunists do best: move into the empty space.

Chapter 6

Wired

We often speak of "gut feelings" without realizing the literal truth behind the metaphor. This chapter traces the physical and chemical cables connecting our intestines to our skull, revealing how microscopic residents hack into the body's control room to influence everything from anxiety to resilience.

6.1 The Hardlines

A person can be lying perfectly still, eyes closed, saying nothing, while their body runs a constant private conversation. The heart adjusts its rhythm. The stomach tightens or relaxes. The intestines squeeze forward a wave of motion that moves dinner along. A brief drop in blood pressure is corrected before it becomes dizziness. None of this waits for conscious permission. The question is not whether the brain is involved, but *how* the body and brain keep in touch moment to moment.

The gut has its own on-site nervous system, dense enough to deserve its famous nickname: the "second brain". The proper term is the *enteric nervous system* (ENS). It is not a vague web of "nerve endings" but a structured network embedded in the wall of the gastrointestinal tract, with ganglia (clusters of nerve cell bodies) and organized circuits. It can coordinate many jobs locally: rhythmic contractions that propel food, the opening and closing of valves, the secretion of acid and enzymes, and the timing of fluid absorption. If the gut were cut off from the skull entirely, much of its choreography would continue. This autonomy is not mystical; it is engineering. When a system must make rapid decisions in an unpredictable environment, it helps to have local control rather than routing every signal to headquarters.

Yet the gut is not a separate republic. It remains part of a larger federation of nerves that monitors internal state and adjusts priorities. The most important "hardline" in that federation is the vagus nerve.

The vagus is sometimes described as a single cable running from brain to belly, like a phone line from a switchboard. Anatomically, it is more like a thick trunk that splits into many branches, and those branches braid into organs that each have their own local wiring. The vagus leaves the brainstem (the part of the brain that merges into the spinal cord) and travels down the neck into the chest and abdomen. Along the way it reaches the heart, the lungs, and much of the digestive tract. Its name comes from the Latin *vagus*, "wandering", and it earns the title. It wanders because it is not devoted to one target; it is a distributed network for managing the body's "rest-and-digest" mode.

A common misunderstanding is that the vagus is mainly a set of instructions from the brain: *Slow the heart. Increase digestion. Calm the immune system.* In reality, the traffic is lopsided in the opposite direction. Roughly 80–90% of vagal fibers are *afferent* (carrying signals from body to brain), while only about 10–20% are *efferent* (carrying commands from brain to body). That imbalance is worth lingering on, because it flips the intuitive story. The vagus is not primarily the brain's leash on the gut. It is primarily the gut's reporting line to the brain.

The distinction between afferent and efferent is a basic one in neuroscience, but it matters here because it reframes what "gut feelings" could mean in literal physiology. Afferent vagal fibers act as internal sensors. They report stretch (how full the stomach is), chemical conditions, inflammatory signals, and the presence of nutrients. They terminate in the brainstem, especially in a hub called the nucleus of the solitary tract. From there, the information is relayed to wider networks that influence mood, attention, motivation, nausea, satiety, and the stress response. Some of that influence is subtle. Some is dramatic. Anyone who has experienced sudden nausea that wipes out the ability to think about anything else has felt how quickly visceral signals can dominate the brain's priorities.

The vagus is not alone, and it does not carry everything. Pain from many gut regions travels primarily by spinal pathways rather than the vagus, and hormones provide a parallel channel, often slower but longer-lasting. Immune signals travel through cytokines and other molecules, and can also be "read" by nerve endings. The point is not that the vagus is the only route, but that it is a privileged one: fast, hardwired, and heavily bottom-up.

A brief historical detour shows how long it took to treat this wiring as a system rather than a curiosity. Galen, writing in the second century, described nerves that influenced voice and swallowing; later anatomists traced branches of a "pneumogastric" nerve (an old term reflecting its reach to lungs and stomach). But for centuries, the prevailing medical imagination leaned toward fluids and vapors: digestion produced miasmas that drifted upward, or "nervous humors" that disturbed the mind. When nineteenth-century physiologists began mapping reflexes with careful experiments, they revealed a different kind of explanation: circuits. A stimulus in the gut could trigger a response without consulting consciousness because it was built into the wiring itself.

One of the classic demonstrations of this circuitry is the family of *vagovagal reflexes*, loops in which vagal afferents report a state of the gut and vagal efferents adjust it. When food enters the stomach, stretch receptors send signals up the vagus. The brainstem integrates that information and sends signals back down to coordinate stomach relaxation, acid secretion, and the timing of emptying into the small intestine. The reflex is not a single on/off switch; it is a continuous feedback controller, the biological equivalent of a thermostat that can anticipate demand.

That brainstem integration occurs in a cluster of nuclei sometimes called the dorsal vagal complex. The name sounds like a building on a university campus, but it is closer to an air-traffic control room for internal state. It is where many visceral signals first arrive, and where output patterns are organized. The gut and the brain are not connected by a poetic metaphor; they are connected by a control architecture.

The ENS and the vagus also meet each other in a way that can mislead. The vagus does not simply "plug into" the ENS as one wire joins another. It interfaces with multiple layers: muscle, glands, immune structures, and specialized sensory cells in the epithelium (the lining). The gut wall is not an inert tube but a living surface exposed to a dense microbial ecosystem. Signals from that ecosystem do not need to be translated into a single chemical that floats into the bloodstream and somehow wafts into the brain. Many signals can be detected at the boundary and converted into nerve activity quickly.

A striking example came from work in the 2010s that challenged an older assumption: that gut sensing reaches the brain mostly through slow endocrine routes. Researchers including Maya Kaelberer and colleagues (2018) showed that certain enteroendocrine cells-cells in the gut lining traditionally framed as hormone-releasing-can form synapse-like connections with vagal neurons. These "neuropod" cells can transmit signals on the timescale of milliseconds, using classical neurotransmission. For a general reader, the technical novelty can be stated simply: the gut lining has cells that behave a bit like sensory neurons, capable of fast point-to-point signaling into the vagus. That helps explain how the brain can react to nutrients with surprising speed, sometimes well before digestion has progressed far enough for meaningful changes in blood chemistry.

Why would evolution build such a fast lane? Because food is not just fuel; it is information. A mouthful of sugar predicts energy availability. Fat predicts caloric density. Bitter compounds predict potential toxins. The gut is the first internal checkpoint where the body can verify what has arrived and decide how to allocate resources: insulin release, stomach emptying, appetite suppression or continuation, and even aversive learning that makes a food unappealing next time. Speed matters when you are making a choice about what to eat in the next few minutes, not the next day.

That same speed and directness are what make the vagus so relevant to the microbiome. The microbes in the colon are mostly physically separated from nerve endings by layers of mucus, epithelial cells, and immune surveillance. They are not chewing on nerves. The influence is more indirect: microbial metabolites, microbial-associated molecular patterns, changes in acidity, and changes in gut barrier function alter what the epithelium and immune system are doing. Those changes can then be reported by sensory fibers. The vagus becomes an interpreter of conditions at the frontier.

It helps to picture the gut wall as a border crossing rather than a simple wall. The epithelium is a single layer of cells, but it is reinforced by tight junctions-protein complexes that seal the gaps between cells-and covered by mucus. Underneath lie immune cells, blood vessels, and nerve networks. When the barrier is intact and calm, microbial signals are largely translated into "normal" physiological language: steady motility, appropriate secretion, and a tuned immune tone. When the barrier is stressed or inflamed, the same microbial environment can be translated into alarm: altered motility, discomfort, and heightened vigilance.

The vagus participates in both reporting and response. Its efferent arm is part of the parasympathetic nervous system, which tends to support digestion, restorative processes, and energy storage. But it also influences inflammation. In experimental contexts, stimulation of vagal pathways can reduce certain inflammatory outputs, a phenomenon often discussed as the "cholinergic anti-inflammatory pathway". The details are complex and still being refined, but the essential idea is clear enough: neural signals can adjust immune activity, not just muscle contraction and gland secretion. This is one reason "vagal tone" has become a popular phrase in wellness circles, sometimes with more confidence than the evidence warrants. The vagus is real anatomy, but it is not a single dial labeled "CALM". It is a bundle of fibers with multiple targets and multiple messages, and the effects depend on context.

A concrete clinical example shows how hardwired reflexes can intrude into daily life. Consider vasovagal syncope-fainting triggered by a sudden drop in heart rate and blood pressure, often in response to pain, emotional distress, or even standing still for too long. The label "vasovagal" reflects the involvement of vagal pathways that slow the heart and vascular responses that reduce blood pressure. For some people, a gut stimulus can be part of the trigger: intense abdominal pain, rapid changes in bowel function, or nausea can precede a faint. This is not a story about "microbes causing fainting". It is a reminder that the vagus links visceral sensation with cardiovascular control, and that signals originating in the abdomen can have whole-body consequences in seconds.

Another familiar experience is the sudden loss of appetite during acute stress. Stress activates sympathetic pathways (the "fight-or-flight" system) and hormonal cascades, and it also changes vagal signaling. In the short term, this can pause digestive investment. In the longer term, if stress becomes chronic, the altered pattern of motility and secretion can reshape the gut environment itself, changing what microbes thrive. The hardlines do not merely transmit messages about the gut; they participate in creating the gut conditions being reported.

The phrase "second brain" can mislead if it suggests a rival mind in the intestines. The ENS does not write poetry or solve algebra. Its intelligence is practical and bodily: coordinating movement and secretion while negotiating with immune defenses and microbial residents. What makes it brain-like is not consciousness but circuit complexity and autonomy. Meanwhile, the brain in the skull is not a detached executive. It is a prediction machine whose job includes keeping the internal environment stable enough for thought and movement. The vagus is one of the key ways the brain learns what is happening inside the body without having to open the abdomen and look.

That learning is not purely informational; it shapes perception and emotion. The brain's representation of internal state is sometimes called *interoception*. If you have ever felt "butterflies" before speaking in public, or a sinking sensation when receiving bad news, you have experienced interoceptive perception as part of emotion. The vagus is central to the raw data of interoception. This makes it tempting to draw a straight line from gut microbes to mood: microbes alter gut chemistry; gut chemistry alters vagal firing; vagal firing alters emotion. The architecture supports the possibility, but biology rarely cooperates with straight lines. The same vagal pathways that can signal satiety can also signal nausea; the same immune signals that can promote recovery can also generate fatigue. The message is not "good" or "bad". It is "state".

One reason the hardlines matter is that they set limits on what kinds of causal claims are plausible. If a proposed gut-to-brain effect would require large amounts of a compound to cross the blood-brain barrier quickly, skepticism is warranted. If, instead, the effect could plausibly occur by changing the activity of sensory fibers at the gut lining, the mechanism becomes more reasonable. This is not a guarantee that any specific claim is true, but it is a way of sorting stories into those that violate anatomy and those that respect it.

The vagus also adds a layer of selectivity. The brain does not receive a full "microbial census". It receives a compressed report: patterns of stretch, nutrient detection, inflammatory tone, and barrier status, integrated into a signal that can adjust appetite, motility, and stress physiology. That compression is important. Most of the microbiome's activity never becomes a conscious sensation. It becomes a background influence on how easily one feels satisfied after a meal, how reactive the gut is to certain foods, how readily nausea is triggered, or how strongly the body responds to a mild immune challenge.

A clean way to think about the hardlines is to imagine two kinds of control happening at once. Locally, the ENS runs the gut's daily operations with exquisite competence. Centrally, the brainstem and higher centers supervise priorities: conserve energy or spend it, speed up transit or slow it, invest in digestion or divert resources to vigilance. The vagus is a main conduit between these levels, and its heavy afferent bias means the gut is not a passive recipient of orders. It is a talkative organ with a direct line to the brain's most ancient regulatory centers.

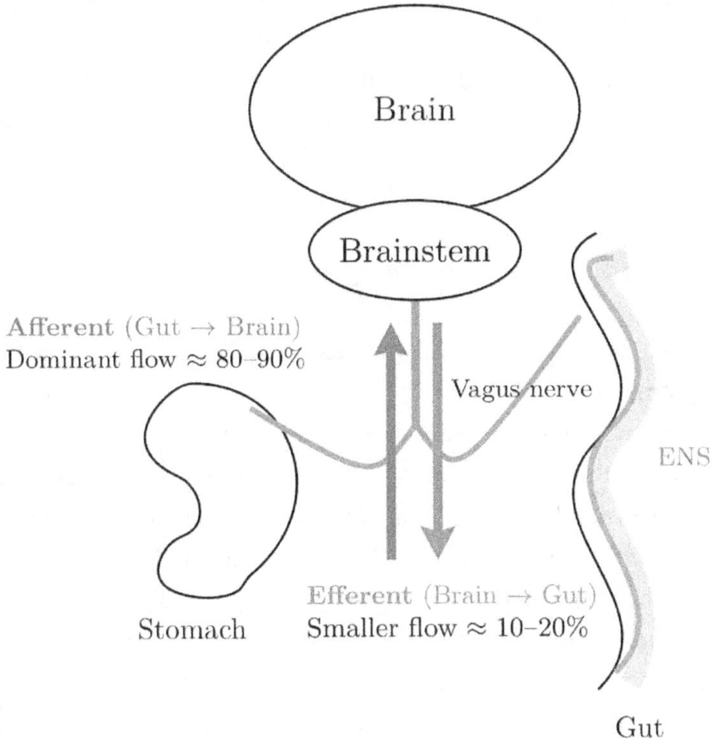

If this anatomy feels oddly relevant to everyday life, it is because the signals it carries are not exotic. They are the simplest facts of being embodied: fullness, nausea, the gentle pressure of gas, the quiet ease after a meal that "sits well", the sudden churn that says something is wrong before a thermometer registers fever. The gut's neurons and the vagus nerve do not translate these states into words. They translate them into priorities. Appetite is postponed or sharpened. Attention narrows. The heart rate shifts. The immune system is nudged. The organism, as a whole, changes what it is ready to do.

The hardlines make one final point unavoidable. Whatever else the microbiome is-an ecosystem, an organ-like community, a source of chemical innovation-it lives in a body that listens. Not always consciously, and not always accurately, but continuously. A nerve the width of a shoelace can carry a stream of information from the gut to the brainstem that never becomes a thought, yet still shapes the tone of the day.

6.2 The Bacterial Pharmacy

Walk into any pharmacy and you meet a particular view of the body: symptoms are problems, chemicals are solutions, and the route from pill to effect is ideally direct. The gut microbiome invites a stranger picture. In the intestine, chemistry is not delivered once or twice a day in measured doses. It is manufactured continuously, in shifting blends, by thousands of species that respond to what you ate, how fast food is moving, whether the lining is inflamed, and which neighbors happen to be thriving. If the nervous system is an electrical network, the microbiome is a chemical industry that runs beside it: fermenting, chopping, reducing, oxidizing, and releasing small molecules into a landscape packed with receptors.

Calling this a "pharmacy" risks the wrong implication, as if the microbes are benevolent pharmacists compounding remedies. Much of microbial chemistry is simply metabolism: making a living. Some products are helpful to the host, some are neutral, and some can be harmful depending on context. The deeper point is that the host is not merely *exposed* to these chemicals. The host is built to *interpret* them. The gut lining is studded with sensors that recognize microbial metabolites, and nerves and immune cells are positioned to respond. Many of the most interesting brain-relevant signals never need to enter the brain at all. They can change gut motility, immune tone, or vagal firing patterns, and the brain responds to those state changes.

The popular imagination tends to fixate on neurotransmitters: serotonin, dopamine, GABA. These are famous because they are associated with mood and anxiety and because modern psychiatry has medications that target their receptors or reuptake. It is true that gut microbes can produce or transform molecules that resemble neurotransmitters, and it is also true that the gut contains the body's largest store of serotonin. But there is a trap here. A neurotransmitter is not magical because of its name; it is powerful because of *where* and *how* it is released. A molecule that acts as a neurotransmitter in a synapse in the brain may behave very differently when it is produced in the colon lumen, separated from nerves by mucus and epithelium, and rapidly metabolized by surrounding cells. The question is not "Do microbes make serotonin?" but "How does microbial chemistry get converted into signals that the nervous system actually uses?"

Serotonin is the most useful place to start, precisely because it is so often misunderstood. In the brain, serotonin (5-hydroxytryptamine, or 5-HT) is a neurotransmitter involved in mood, sleep, appetite, and many other functions. Many antidepressants, including selective serotonin reuptake inhibitors (SSRIs), work by changing how long serotonin remains available at synapses. It is tempting to assume that if the gut produces serotonin, it must be sending serotonin to the brain. That is not how the plumbing works. Most serotonin in the body is made in the gut, largely by specialized epithelial cells called *enterochromaffin cells*. Much of it acts locally: shaping gut movement and secretion. Some enters the bloodstream, where platelets take it up and carry it, influencing processes like clotting. But serotonin itself does not freely flow into the brain in large amounts because the blood—brain barrier is selective. This barrier is not a wall but a living interface of tightly joined endothelial cells, transporters, and metabolic enzymes that strongly regulate what reaches neural tissue.

This is where the microbiome becomes interesting in a more subtle way. Microbes may not need to deliver serotonin to the brain if they can influence *how much serotonin the host makes* in the gut, or how enterochromaffin cells behave, or how nerves in the gut wall respond to what the epithelium releases. Jessica Yano and Sarkis Mazmanian's group (2015) provided a landmark demonstration that indigenous gut bacteria can regulate host serotonin biosynthesis. The key is not that the bacteria are tiny serotonin factories that replace the host's chemistry, but that they are signals in an ecosystem that tunes the host's own output. When you change the microbial community, you can change the host's serotonin levels in the gut, with consequences for motility and platelet function. In other words, microbial presence becomes part of the body's baseline settings.

A historical echo helps here. In the mid-twentieth century, serotonin was first isolated and linked to blood vessel tone; its very name comes from "serum" and "tone." Only later did it become famous as a brain neurotransmitter and then as the target of psychiatric drugs. Serotonin has always been a molecule with multiple lives, depending on location. The gut did not borrow serotonin from the brain; the brain is one of serotonin's many arenas. The microbiome's influence, when it is real, often comes through altering those arenas indirectly.

GABA (gamma-aminobutyric acid) creates a different kind of temptation. In the brain, GABA is the major inhibitory neurotransmitter: it helps dampen neuronal firing and is central to anxiety circuits. Many sedatives and anti-anxiety drugs ultimately increase GABA signaling. Certain bacteria can produce GABA from glutamate, and probiotic marketing has seized on this fact with predictable enthusiasm. But again, the basic constraint remains: GABA made in the gut lumen is not the same as GABA released at a synapse in the amygdala. The more serious question is whether microbial activity can change *host* GABAergic signaling in the brain through plausible routes: vagal afferents, immune mediators, endocrine changes, or precursor availability.

One of the most widely discussed experiments linking a specific microbe to brain GABA signaling is the study by Javier Bravo, Paul Forsythe, and colleagues (2011). In mice, ingestion of a particular *Lactobacillus* strain was associated with changes in central GABA receptor expression and with altered stress-related behavior, alongside changes in stress hormones. The striking mechanistic twist was that the effect disappeared when the vagus nerve was cut. That single detail is more valuable than a dozen breathless headlines: it suggests a route by which a gut perturbation can influence brain chemistry without requiring microbial metabolites to pour into the brain. The message, in that animal model, traveled via the hardlines.

Animal studies like this are not proof that a yogurt will treat anxiety in humans. They are proof that the body contains wiring and interfaces that make such influences *possible* under some conditions. Translating that possibility to people is difficult: our microbiomes are more diverse, our diets and stressors are less controlled, and our mental states are not measured by a single standardized behavior in a maze. Still, the experiment remains a clarifying example: microbial signals can be turned into neural messages, and those messages can reshape receptor expression in the brain.

If classic neurotransmitters are the celebrities, short-chain fatty acids are the working-class majority. These small molecules—acetate, propionate, and butyrate—are produced when microbes ferment dietary fiber and resistant starch, the carbohydrates our own enzymes cannot fully digest. For many readers, “fiber” is a bland nutritional slogan. In microbial terms, it is a primary fuel source. Without fermentable substrates, many beneficial microbes cannot maintain their populations, and the chemical environment of the colon changes.

Short-chain fatty acids (SCFAs) matter for gut health because colon cells use butyrate as an energy source, and SCFAs help maintain barrier integrity and modulate immune responses. The brain-related question is how SCFAs could influence the nervous system. Several routes are plausible, and the current evidence suggests a mixture rather than a single pipeline. SCFAs can enter the bloodstream, reaching peripheral organs and potentially affecting the brain's vascular and immune interfaces. They can act on receptors expressed on immune cells and endocrine cells. They can influence gene expression through epigenetic mechanisms such as inhibiting histone deacetylases (HDACs), which changes how tightly DNA is packaged and how genes are transcribed. They can also act locally in the gut to change motility and secretion, altering the pattern of vagal afferent signaling.

Kenneth O'Riordan and colleagues (2022) summarized the field's growing attention to SCFAs as candidate mediators of microbiome—brain communication. Yet SCFAs also illustrate why the word "pharmacy" needs a warning label. Dose and context matter. A molecule that supports barrier function in one setting may contribute to symptoms in another. A receptor that mediates anti-inflammatory effects in one tissue may have different consequences elsewhere. Even the same SCFA can be protective or problematic depending on whether the gut lining is intact, whether the immune system is primed, and which other metabolites are present.

Tryptophan is another key node, partly because it sits at the crossroads between diet, microbes, and the host's own neurotransmitter and immune pathways. Tryptophan is an essential amino acid; you must obtain it from food. The host uses it to make serotonin, but tryptophan can also be metabolized along other routes, including the kynurenine pathway, which is closely tied to immune activation. Microbes, for their part, can transform tryptophan into a variety of indole compounds. These indoles are not neurotransmitters in the classic synaptic sense, but they can act as signaling molecules by binding to host receptors.

One receptor that has emerged as a particularly interesting "listening post" is the aryl hydrocarbon receptor (AhR). It is found in many cell types, including immune cells and barrier tissues, and it responds to a range of ligands, including some derived from microbes. Researchers in the 2020s used germ-free animals and microbial gene knockouts to show that specific microbial tryptophan metabolism can alter host physiology through AhR signaling, with downstream effects that reach into the nervous system in animal models. The appeal of this line of work is not that it offers a single miracle metabolite, but that it offers a credible chain of causation: microbial genes produce a defined chemical; that chemical binds a defined host receptor; and the host's response can be measured.

Bile acids provide yet another example of how the microbiome behaves like a chemical processor rather than a mere passenger. The liver makes primary bile acids from cholesterol and secretes them into the intestine to help digest fats. Microbes then modify these bile acids into secondary forms. These molecules are not only detergents; they are signals. They can bind receptors such as FXR and TGR5 that influence metabolism and inflammation. Because metabolic and inflammatory states affect the brain, bile acid signaling becomes another plausible route by which microbial activity could influence neurobiology. The complexity here is formidable: bile acid species differ in their receptor affinities and biological effects, and the balance among them can shift with diet, antibiotics, disease, and microbial composition. Some bile acid patterns appear neuroprotective in certain contexts and neurotoxic in others. The same chemical family can play both roles, depending on the ecology that produces it and the host state that receives it.

At this point, it helps to step back and ask what sort of evidence deserves confidence. In human studies, researchers often measure microbial taxa (which species are present) and correlate them with symptoms like anxiety, depression, or cognitive complaints. These studies are valuable for mapping patterns, but correlations can be misleading. People who are depressed may eat differently, sleep differently, exercise less, take different medications, and experience different stress hormones, all of which can shift the microbiome. The direction of causality can be circular or reversed. Even when microbial differences are consistent, the active agents may not be the microbes themselves but what they are doing metabolically. Two different microbial communities can produce similar SCFA profiles, or similar bile acid transformations, depending on diet and other conditions.

Mechanistic studies that move beyond correlation often use germ-free animals (raised without microbes) or antibiotics to deplete microbes, then reintroduce defined communities. These approaches can show that microbes are *necessary* for certain host phenotypes and can sometimes show that a particular strain or metabolite is *sufficient* to reproduce an effect. But animals are not small humans. Their diets are standardized, their genetics are inbred, and their environments are controlled. A microbial product that shifts a mouse's stress behavior may not produce a noticeable change in a human mind embedded in social life.

Even within the gut, the geography matters. The small intestine is relatively oxygenated and fast-moving compared with the colon; it hosts fewer microbes. The colon is denser, slower, more anaerobic, and more fermentation-heavy. A metabolite produced in the colon may act mostly there, while the vagal sensory endings that respond most strongly might be in other regions, or might respond indirectly through enteroendocrine signals. The idea of a single "gut chemical" that travels to the brain is usually too simple. The more realistic picture is a layered translation: microbes alter the chemical environment; the gut epithelium and immune system translate that environment into host signals; nerves and hormones carry those signals onward; and the brain integrates them with everything else it knows.

Concrete examples help keep this grounded. Consider what happens when someone switches abruptly from a low-fiber diet to a high-fiber diet heavy in legumes, whole grains, and certain vegetables. Many people experience gas, bloating, changes in stool consistency, and altered appetite for a period of adaptation. The easy explanation is mechanical: "fiber adds bulk." The microbial explanation is chemical and ecological: new substrates arrive, fermentation patterns shift, SCFA production changes, and gas production changes. The gut wall senses these shifts through stretch, chemical receptors, and immune signals. The nervous system responds by adjusting motility and sensitivity. Over time, the microbial community can adapt, symptoms often improve, and the person may notice a steadier appetite or different satiety patterns. This is not a guaranteed mental health intervention; it is a lived demonstration that microbial metabolism can alter bodily signals that shape mood and attention indirectly.

A different example comes from antibiotics. Many people have experienced an unsettling "off" feeling during or after a course of broad-spectrum antibiotics: not just diarrhea or stomach discomfort, but fatigue, irritability, sleep disturbance, or brain fog. These experiences are non-specific and can have multiple causes, including the infection itself. But they are consistent with an ecosystem shock. When microbial populations are abruptly reduced, the chemical output of the gut changes: less fermentation, altered bile acid metabolism, altered production of microbial-derived vitamins and metabolites. The gut barrier can become temporarily stressed. Immune tone can shift. The brain, which is exquisitely sensitive to systemic inflammation and metabolic state, may register this as malaise or cognitive drag. Here again, the story is not that microbes "secrete happiness." It is that the microbiome helps maintain a baseline chemical environment that supports normal physiology, and sudden perturbations can reverberate.

The phrase "psychobiotics" has been used to describe microbes (or microbial-targeted interventions) that might improve mental health. It is an appealing idea, and it may eventually yield useful therapies for some conditions. But the best current science suggests a more cautious framing. Microbial chemistry interacts with the nervous system most reliably through general mechanisms: barrier integrity, immune modulation, metabolic signaling, and vagal pathways that report internal state. These mechanisms can influence mood and cognition, but they do so as part of whole-body regulation, not as a direct pharmacological replacement for a psychiatric drug.

A useful way to hold the nuance is to keep asking a simple question: *What is the receptor, and where is it?* If a microbial metabolite binds a receptor on an enteroendocrine cell, that cell can release host hormones or neurotransmitters locally, and nerves can detect that. If a metabolite binds a receptor on an immune cell, cytokine profiles can change, and the brain can respond to inflammation. If a metabolite reaches the bloodstream, it may act on the liver, adipose tissue, or vascular endothelium, all of which affect brain function indirectly. Only a subset of molecules will reach the brain in meaningful concentrations, and even then, the brain's response depends on state: sleep deprivation, chronic stress, infection, adolescence, aging.

There is a quiet, almost philosophical implication in all this chemistry. The microbiome does not simply add extra molecules to the body. It changes the meaning of food. A bean is not just a collection of calories and protein; it is a set of polysaccharides that become SCFAs only in the presence of the right microbial machinery. The same meal can generate different chemical messages in two different guts. That variability makes microbiome science hard, but it also makes it personal. The "pharmacy" inside you is not stocked with identical bottles on identical shelves. It is a living supply chain that depends on what you feed it, what has happened to it, and what your body asks of it.

If the brain is waiting for a single microbial molecule to solve sadness, it will be disappointed. If it is listening for the ongoing report of internal state—whether the gut is calm or inflamed, whether the barrier is secure, whether nutrients are arriving predictably, whether microbial fermentation is producing supportive metabolites or irritating byproducts—then microbial chemistry becomes less like a miracle cure and more like a steady hand on the body's homeostatic controls. That kind of influence is not theatrical, but it is real enough to shape how it feels to live inside your own skin.

6.3 The Stress Loop

Stress is usually described as something that happens in the mind: a deadline, an argument, a sleepless night of worrying. The body tells a different story. Stress is a physiological state that reallocates resources, shifting blood flow, changing immune priorities, and altering how the gut moves and seals itself. The microbiome lives inside that state. When stress becomes recurrent rather than brief, the microbial ecosystem is not a bystander. It is pushed, reshaped, and sometimes destabilized. Then the altered ecosystem can push back, amplifying the very stress physiology that changed it in the first place.

The language of "fight-or-flight" is familiar, but it is only half the stress response. Alongside the fast wiring of the sympathetic nervous system runs a slower endocrine cascade known as the hypothalamic—pituitary—adrenal axis, usually shortened to the

HPA axis. The hypothalamus is a brain region that monitors internal state and threat; the pituitary is a small gland that releases hormonal signals; the adrenal glands sit on top of the kidneys. When the HPA axis activates, it leads to the release of glucocorticoids, chiefly cortisol in humans. Cortisol is not a villain. It helps mobilize energy, regulates inflammation, and tunes memory and attention. The trouble comes when a system designed for acute challenges is held in the "on" position for weeks or months.

Stress reaches the gut quickly. Sympathetic activity can slow digestion, alter secretion, and reduce blood flow to the intestinal lining. Cortisol and related hormones can change how immune cells behave in the gut wall. The vagus nerve, which carries a heavy stream of sensory traffic from gut to brainstem, also changes its pattern under stress; "vagal tone," a rough term for the vagus's influence on organs like the heart, often decreases during sustained stress. A gut that is less well perfused, more inflamed, and less well regulated is a different habitat. Microbes are exquisitely sensitive to habitat. They respond to oxygen gradients, bile acids, mucus composition, available carbohydrates, and the timing of intestinal transit. Stress alters all of these at once.

A useful metaphor here is ecological, not psychological. Think of the colon as a rainforest. There is a stable climate, a predictable day-night rhythm of food arriving, and a complex web of species that keep each other in check. Now imagine a long drought. Some plants die back, others take over, and opportunistic species expand into newly opened niches. When the rains return, the forest may recover, or it may settle into a different equilibrium. The gut ecosystem can behave similarly. Stress is not a single toxin; it is a climate shift.

One of the most discussed physiological pathways is the effect of stress on the gut barrier. The gut lining is only one cell thick, yet it normally does an extraordinary job: it absorbs nutrients while keeping most microbes and their components at a safe distance. Tight junctions between epithelial cells help maintain this boundary. Mucus provides an additional layer, and immune cells patrol beneath. Under stress, tight junctions can become less effective, mucus properties can change, and inflammation can increase. The result is often described, somewhat crudely, as "leaky gut," meaning increased intestinal permeability. It does not mean the gut becomes an open sewer. It means that the filtering function becomes less selective. Small amounts of bacterial products that would normally stay contained can cross into the body.

The bacterial products most often invoked are fragments of cell walls, including lipopolysaccharide (LPS) from gram-negative bacteria. LPS is a potent immune trigger. If it reaches the bloodstream in meaningful amounts, it can provoke systemic inflammation. Inflammation, in turn, has well-established effects on the brain: it can alter sleep, reduce motivation, increase pain sensitivity, and produce a heavy fatigue sometimes called "sickness behavior." These are adaptive responses during infection. In chronic low-grade inflammation, they can resemble depression or anxiety.

The idea that gut-derived immune activation might contribute to some forms of depression has a long history in modern psychoneuroimmunology. Clinical studies have reported, in some cohorts of depressed patients, increased immune responses (including IgA and IgM antibodies) to LPS from gut commensals, interpreted as consistent with increased bacterial translocation. The wording matters. "Consistent with" is not "proven." These biomarkers vary across studies, and measuring permeability in humans is notoriously difficult. Different methods capture different aspects of barrier function, and the gut is a long organ with regional differences. Still, the recurring pattern is suggestive enough to take seriously: in at least some people, chronic stress and depression-like physiology coincide with signs of a stressed gut barrier and heightened immune reactivity to microbial components.

The most convincing evidence for causation often comes from animal models where stress can be imposed in a controlled way and the microbial ecosystem can be measured before and after. In rodents, chronic stress paradigms can shift microbial composition, reduce microbial diversity, and increase markers of inflammation and permeability. When the microbiome is experimentally manipulated—for example, by transferring microbes from stressed animals to unstressed ones—some stress-related behaviors and immune patterns can partially transfer as well. These experiments are not perfect mirrors of human life, but they demonstrate the existence of a loop: stress changes the gut and its microbiome; the altered microbiome changes host physiology in ways that resemble stress.

A human-scale example makes the loop more relatable. Consider the period around major exams for students, or the weeks around a grueling work deadline. Many people report gastrointestinal symptoms: constipation, diarrhea, cramping, nausea, reflux. These are not imagined; they are physiological outputs of altered motility, secretion, and sensitivity. During such periods, diet often changes (more ultraprocessed food, less fiber), sleep shortens, and alcohol or caffeine intake increases. Each of those factors can shift the microbiome. The microbial changes are not always dramatic enough to show up as a single signature across people, but the *direction* of pressure is similar: less consistent feeding of fiber-fermenting communities, more inflammatory tone, and often less stable daily rhythms. Then, as GI symptoms increase, the person may become more vigilant about bodily sensations, more anxious about leaving the house, or more avoidant of meals that "might cause problems." The brain interprets the gut's signals as threat, and threat physiology further perturbs the gut. What began as a psychological stressor becomes a whole-body feedback loop.

A historical anecdote sits in the background of much of this work: the discovery that the immune system can generate an internal state that looks like depression. In the late twentieth century, clinicians observed that patients receiving cytokine therapies (for example, interferon treatments) often developed profound fatigue and depressive symptoms. This was not a moral weakness or a mere reaction to diagnosis; it was biology. Inflammation can change mood. That insight made it easier to imagine that chronic, low-grade inflammatory signaling from the gut could alter the brain's affective circuits. It also helped researchers frame depression not as a single disease but as a family of syndromes with multiple biological routes.

The vagus nerve sits at an intersection of these routes. It can carry information about gut inflammation to the brain, and it can also participate in regulating inflammation. Experimental work on the "cholinergic anti-inflammatory pathway" suggests that activating vagal efferents can dampen pro-inflammatory cytokine release through mechanisms involving acetylcholine and receptors such as the α_7 nicotinic acetylcholine receptor. This is a striking concept: the nervous system can, in effect, apply a brake to immune overreaction. It has motivated interest in vagus nerve stimulation as a therapeutic approach for inflammatory disorders, and has been discussed as a potential route by which stress resilience might be supported physiologically.

It is also a reminder that stress is not simply "too much cortisol." Stress is a shift in balance among multiple control systems. A person with low vagal tone and heightened sympathetic arousal may have a gut that is more reactive and an immune system that is more easily nudged toward inflammation. In that state, microbial perturbations may have larger effects, and inflammatory messages may reach the brain more readily.

The loop becomes more concrete when viewed through the lens of timing. Microbes respond quickly to changes in diet and transit time. Some changes in microbial gene expression can occur within hours. Barrier changes can also happen quickly. The brain's stress response, particularly the sympathetic arm, can shift within seconds. That means the loop can operate on multiple timescales at once: a fast loop where anxiety triggers gut sensations that trigger more anxiety, and a slower loop where weeks of poor sleep and altered eating patterns gradually reshape the microbiome and inflammatory tone, making the fast loop easier to trigger.

The Bravo and Forsythe mouse study (2011) mentioned earlier offers a useful example of how these loops can be experimentally probed. When the vagus nerve was cut, the probiotic strain no longer produced its observed effects on stress-induced corticosterone (a rodent glucocorticoid) or on central GABA receptor expression and behavior. The result does not mean "the vagus controls emotions." It means that for at least one microbial perturbation in one animal model, the vagus was a necessary conduit. In the context of stress loops, this supports the idea that microbial changes can influence the brain not only through circulating molecules but through sensory wiring that reports gut state.

This is where caution is essential. In popular discourse, "gut-brain axis" sometimes becomes a license to claim that anxiety is simply a microbiome imbalance waiting for a supplement. That does a disservice both to patients and to the science. Stress-related conditions are multifactorial. A microbial shift may contribute to symptoms in some people, while in others it may be a downstream marker of altered sleep, diet, or medications. The same microbial feature may be protective in one context and harmful in another. The most defensible claim is narrower: the gut ecosystem participates in the physiology of stress, and under certain conditions it can amplify or dampen the body's stress response.

One reason the loop is hard to break is that many stress-reduction strategies act indirectly on the microbiome, while many microbiome-targeted strategies act indirectly on stress. Consider diet. Increasing dietary fiber can increase SCFA production, which tends to support barrier function and anti-inflammatory immune patterns, and can influence satiety and metabolic stability. Those changes may make the body less reactive to stress. But the dietary change is itself hard to maintain under stress, when appetite shifts toward convenience foods and the gut may be more sensitive to fermentable fibers. Similarly, exercise can reduce stress and modulate inflammation, and it also tends to increase microbial diversity in some studies, but exercise is often the first habit to disappear during burnout. Sleep is another lever: poor sleep increases stress hormones and inflammation and is associated with microbial shifts; improving sleep can support both brain and gut, yet sleep is easily disrupted by anxiety and gastrointestinal discomfort.

These practical observations highlight a deeper scientific challenge: causality is braided. The most straightforward causal experiments in humans are difficult or unethical. You cannot randomize people to chronic stress. You cannot readily cut the vagus nerve. Even "simple" interventions like probiotics or diet changes are hard to blind, and outcomes like mood are sensitive to expectations. This is why biomarkers matter and why they also frustrate. Measures of permeability, LPS-related immune responses, inflammatory cytokines, and microbial metabolites can help triangulate mechanisms, but they are noisy. Different laboratories use different assays. Different cohorts differ in medication use, obesity rates, sleep quality, and socioeconomic stress. A result that replicates in one setting may fail in another without necessarily being wrong.

Despite these difficulties, the stress loop has an emotional resonance that can be clinically useful, even before the science is fully settled, because it reframes symptoms in a compassionate way. Many people with chronic anxiety or depression report gastrointestinal complaints and feel dismissed: told it is "just stress," as if stress were imaginary. The biology suggests the opposite. Stress is not "just" anything. Stress is a whole-body state that can alter the gut's ecology and barrier function, and those changes can feed back into the brain via immune and neural pathways. Feeling trapped in a loop is not a personal failure. It is a property of coupled control systems.

The most hopeful implication is not that there is a single button to press, but that loops have more than one point of entry. A small improvement in sleep can reduce cortisol and sympathetic tone, which can improve motility and barrier function, which can make the microbiome's habitat more stable, which can reduce inflammatory signaling, which can make sleep easier. Conversely, a gentle dietary change that supports fermentation and barrier integrity can reduce gut discomfort and improve metabolic stability, which can blunt some stress reactivity and make psychological coping skills more effective. The loop can run in either direction.

A final image helps fix the idea. Imagine the brain as a weather station that forecasts danger. It does not only look outward; it also reads internal signals. If the gut is inflamed, permeable, or chemically unsettled, the brain receives a steady drizzle of warnings: altered vagal input, immune mediators, sleep disruption, appetite changes. Under that drizzle, the brain's forecasting becomes more pessimistic. The world feels sharper-edged. Small challenges feel bigger. The forecast drives behavior, behavior reshapes physiology, physiology reshapes the gut ecosystem, and the ecosystem reshapes the internal signals that the brain uses to forecast. The loop is not fate, but it is real, and its existence makes the body's quiet messages worth listening to with both curiosity and care.

Chapter 7

Ecosystem Collapse

When the invisible rainforest within us loses its balance, the consequences ripple through our entire physiology. We track the transition from symbiosis to pathology, discovering how our own passengers can drive obesity, ignite chronic inflammation, or succumb to deadly opportunists when their defenses are breached.

7.1 The Energy Harvest

A useful way to picture the gut microbiome is as a living chemical plant bolted onto the side of human digestion. The stomach and small intestine do the obvious work: acids, enzymes, bile, and an assembly line of transporters that pull sugars, amino acids, and fats into the bloodstream. Yet a surprising amount of what we call "food" is really *feedstock* for microbes. Many plant carbohydrates reach the colon largely intact: certain fibers, resistant starches, and complex polysaccharides that human enzymes cannot dismantle. There, in the low-oxygen world of the large intestine, bacteria and archaea take over, fermenting these leftovers into small molecules that *we* can absorb.

That microbial after-dinner labor is one reason the microbiome can be thought of as an organ: it has a steady metabolic output, it communicates chemically with the host, and it differs from person to person in ways that matter. When the system is stable, the arrangement is mostly mutualistic. The microbes get a warm, wet habitat and a constant supply of substrates; the host gets vitamins, a calmer immune tone, and an extra stream of energy in the form of short-chain fatty acids (SCFAs) such as acetate, propionate, and butyrate. But the very fact that microbes can convert "indigestible" material into absorbable calories raises an uncomfortable possibility: if two people eat the same meal, could one microbiome squeeze out more usable energy than the other?

That question moved from speculation to experiment in the early 2000s, when germ-free mice became a kind of blank canvas for microbiology. These animals are raised without any microbes at all. Their intestines are anatomically and immunologically unusual, and they require careful husbandry, but they offer a clean way to test what happens when a particular microbial community is introduced into an otherwise identical host. In Jeffrey Gordon's lab, this approach produced an arresting result: transplanting gut microbes from obese donors into germ-free mice could make the recipients gain more fat than mice receiving microbes from lean donors, even when the animals were not being force-fed and were otherwise comparable. A key 2006 study by Peter Turnbaugh, Ruth Ley and colleagues framed this as an "obesity-associated" microbiome with an increased capacity for energy harvest, shifting the conversation from *which* foods cause obesity to *how* foods are processed by a human—microbial partnership.

The experiment was memorable because it felt almost like a fable. Two cages of genetically similar mice, same chow, same light cycle. The difference was invisible: an introduced community of microbes. Yet the bodies of the animals began to diverge. For readers accustomed to hearing obesity discussed strictly in terms of willpower, calories, or even single genes, the implication was unsettling. Body mass could be nudged by ecology.

The story sharpened years later when Vanessa Ridaura, Joshua Faith and colleagues (2013) used a cleverer human link: they studied pairs of twins in which one twin was obese and the other lean. Twins help control for genetics and many early-life exposures, and they provide a natural "matched pair" design. When microbiota cultured from these discordant twins were transferred into germ-free mice, the recipients of the obese-donor communities tended toward greater adiposity than those receiving the lean-donor communities. Even more intriguing, when mice carrying the two different communities were housed together, the lean-associated community could partially invade the obese-associated one, nudging the phenotype toward leanness—but only under certain dietary conditions. In other words, the microbiome was not a simple pathogen that "caused" obesity; it was an ecological state that interacted with diet, competition, and opportunity.

The first temptation is to translate this into a headline: *obesity is contagious*. But the responsible interpretation is subtler. These are controlled experiments in mice, often using germ-free animals with a distinctive physiology. They show that gut microbial communities can *contribute* to host metabolism in a causal way. They do not prove that obesity spreads between humans like a cold. Still, the work changed the tone of the field. Obesity could no longer be treated as a purely human equation of intake and expenditure. The microbiome entered the ledger.

The mechanism begins with fermentation. When microbes break down complex carbohydrates, they produce SCFAs. These molecules are small enough to be absorbed across the colon lining and used by the host. Butyrate is a major fuel for colon cells themselves; propionate travels to the liver and can be used in gluconeogenesis; acetate circulates widely and can enter lipid synthesis pathways. From a strictly caloric perspective, this is extra energy extracted from food that would otherwise pass through. It is not enormous in every context, but it is not trivial either, and it may matter most over long time scales, when small daily differences compound.

Yet SCFAs are not merely calories. They behave like signals. Reviews by Harry Flint and Karen Scott (2012) and later mechanistic work summarized by Ara Koh, Filipe De Vadder and colleagues (2016) emphasized that SCFAs can activate receptors on gut and immune cells and can influence gene expression through effects on histone deacetylases. Translated into everyday terms, the products of microbial fermentation can act like messages that tune appetite, gut motility, inflammation, and the way fat is stored. The gut is not just absorbing nutrients; it is listening to them.

Consider the gut's endocrine system. The intestine is laced with specialized cells that release hormones in response to what arrives in the lumen. Some of these hormones affect satiety and insulin dynamics. SCFAs can stimulate certain gut hormones through receptor-mediated pathways, influencing how quickly the stomach empties and how full a person feels after a meal. In animal models and human studies, these effects are complex and sometimes contradictory because they depend on which SCFAs are produced, where, and in what amounts, as well as on the host's baseline physiology. But the larger point holds: microbial metabolism can reach beyond the colon, shaping behavior and metabolism upstream.

There is a second, less intuitive pathway by which a microbial community can change the host's energy balance: it can alter what substrates are available to *other* microbes and to the host. A community is not a list of species; it is a network of trades. One microbe breaks down a complex plant polymer into smaller sugars; another consumes those sugars and releases acids; a third uses those acids and produces something else. This cross-feeding means that the capacity for energy harvest is distributed. A gut ecosystem with a dense network of carbohydrate-active enzymes can strip a diet for usable fragments more effectively than one without. That was one reason Turnbaugh and colleagues emphasized metagenomics—the gene repertoire—rather than taxonomy alone. Who is present matters, but so does what their collective genome can do.

This brings us to an idea that briefly became famous and then, in some circles, infamous: the ratio of two major bacterial groups, Firmicutes and Bacteroidetes. Early studies reported that obese individuals, or obese mice, sometimes had more Firmicutes and fewer Bacteroidetes, and that this shift correlated with greater energy harvest. The ratio was attractive because it was simple. It offered a microbiome equivalent of cholesterol numbers: a tidy readout that might predict risk.

But the gut rarely rewards tidy stories. As more cohorts were studied across countries, ages, diets, and sequencing methods, the Firmicutes/Bacteroidetes pattern proved inconsistent. Some studies found it; others did not; some found the opposite; many found that the most informative differences were not at the level of broad phyla at all, but at finer taxonomic resolution, or in functional genes, or in metabolites. The episode is instructive, not embarrassing. It shows how easy it is to mistake a rough early signal—real in a particular experimental context—for a universal biomarker. Ecological systems rarely have a single knob.

A more durable concept is *efficiency*. Two gut ecosystems might produce different amounts of SCFAs from the same amount of fiber. They might differ in how much gas versus acid they generate, in how quickly they ferment, in where along the colon the fermentation happens, and in whether the products favor microbial growth, host absorption, or excretion. Even when the same molecules are produced, they can have different effects depending on timing and location. Butyrate produced close to the colon lining may nourish barrier cells; acetate produced in excess and absorbed rapidly might contribute to lipogenesis under some conditions. Biology turns on context.

A concrete historical anecdote makes this less abstract. Long before microbiome sequencing, physicians observed that antibiotics could have metabolic side effects. In agriculture, low-dose antibiotics were famously used to promote growth in livestock, a practice that later raised serious concerns about antibiotic resistance. The agricultural observation did not prove a microbiome mechanism, but it hinted at one: altering gut bacteria could alter weight gain. Decades later, with modern tools, the idea became testable in a mechanistic way. The livestock story is not a blueprint for human obesity, but it is a reminder that we have been running microbial experiments on animals (and, inadvertently, on ourselves) for a long time.

Diet complicates the picture further because it is not one variable. A diet high in refined sugars and fats can reduce the amount of complex carbohydrate reaching the colon while also changing bile acid flow and gut transit. A diet rich in diverse fibers feeds different guilds of microbes and tends to increase the production of certain SCFAs, often with beneficial effects on the gut barrier and immune tone. But the same high-fiber intervention can have different outcomes in different people because their baseline microbial communities differ. Some people have abundant fiber-degrading microbes ready to respond; others may not. The microbiome is not a passive recipient of diet; it is a filter that determines what “diet” means at the molecular level.

One of the most striking implications of the twin-to-mouse work was the role of competition. When lean- and obese-associated communities were put in contact through co-housing, the lean-associated microbes could sometimes invade and shift the community. But invasion was not guaranteed. Diet could act like border policy: it could open niches for newcomers or lock them out. This is an ecological principle dressed in biological clothing. If you want a new species to establish in an ecosystem, you need to offer it food and space, and you need to consider the defenses of the incumbents.

The popular version of this idea is often phrased as "add probiotics" or "take a supplement." But the deeper message is more demanding and, arguably, more empowering: microbial communities respond to *environments*. Food is the main environment we control daily. When we change that environment persistently, we change which organisms thrive, which pathways are active, and which metabolites are produced.

Obesity and metabolic syndrome are not just about fat storage; they are about how the body handles glucose, how the liver packages lipids, how muscle responds to insulin, and how inflammation smolders in tissues. The microbiome can plausibly influence all of these through its metabolic outputs and its interactions with the host immune system. Even modest shifts in gut permeability or immune activation can affect insulin sensitivity. Even small changes in satiety signaling can affect intake. And even small increases in energy harvest, if consistent, can contribute to weight gain over years.

The danger is to misunderstand this as fate written by bacteria. Microbiomes are plastic, especially when the environment changes. Yet they are not infinitely malleable, and they do not all respond in the same direction. That is why simplistic prescriptions—one microbe to add, one ratio to optimize, one forbidden food—tend to disappoint. An ecosystem can be nudged, but it cannot be micromanaged without consequences, and it rarely snaps into a new stable state overnight.

It helps to return to the dinner plate. Imagine two people eating a bowl of beans and vegetables. On paper, the meal contains a similar number of calories for both. But in one colon, a dense consortium of fiber degraders rapidly breaks down complex polysaccharides, producing a steady stream of acetate and propionate that are absorbed and that stimulate certain gut hormone patterns; in the other, fermentation is slower, less complete, and produces a different balance of metabolites. Over an afternoon, the first person may extract a bit more energy and feel hunger differently; over months, this could subtly influence weight and glucose control, especially when paired with modern patterns of sedentary life and calorie-dense foods. No single meal determines the outcome. But the microbial "energy harvest" turns the gut into a variable component of metabolism rather than a fixed pipe.

This perspective also reframes certain cultural arguments about body size. It does not absolve personal agency, and it should not be used to stigmatize anyone as "microbially broken." Instead, it expands the list of forces acting on the body, including forces we did not evolve to experience at modern intensity: ultra-processed diets low in fermentable substrates, repeated antibiotic exposures, and sanitized environments that change early-life colonization. If the gut microbiome is part of the metabolic apparatus, then it is also part of the metabolic vulnerability.

Perhaps the most practical takeaway is also the least glamorous: a resilient microbial ecosystem is usually one that is well fed in the ecological sense. Diverse, consistent inputs of fermentable plant material tend to support a more diverse community and a richer SCFA profile, though individual responses vary. A microbiome that is starved of its preferred substrates may turn to other resources, including the host's own mucus layer, and may become less stable. Stability is not a moral virtue; it is a property of ecosystems that determines how they behave under stress.

The energy harvest, then, is not a trick played by bacteria on an unsuspecting host. It is the ancient logic of symbiosis: another organism performs chemistry we cannot, and we pay it with habitat and food. In a world of scarcity, this partnership is an advantage. In a world of abundance, especially abundance of foods designed to be rapidly absorbed and easy to overconsume, the same partnership can tilt the scales. The gut microbiome does not replace the familiar stories of calories, exercise, genetics, and social conditions; it threads through them, changing how the body experiences the same environment.

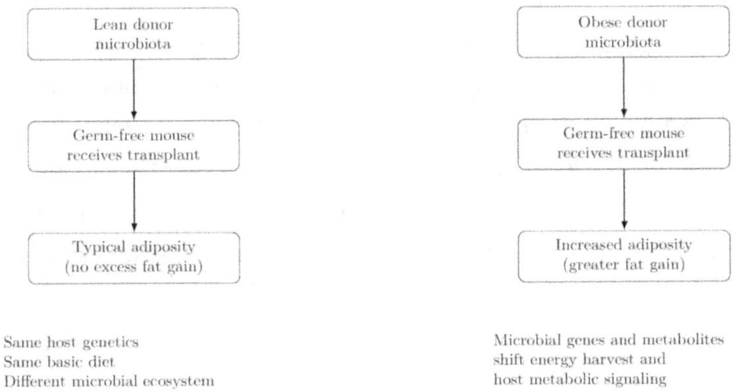

The image is simple, but the implication is profound: a transplanted community can carry metabolic tendencies with it. Not destiny, not a single switch, but a bias—an ecological push that becomes visible when food, physiology, and time provide the right conditions.

7.2 Fire in the Gut

On the surface, Inflammatory Bowel Disease sounds like a problem of *too much* immunity: a body that cannot stop attacking. But the more revealing picture is ecological. The intestine is the most densely populated habitat the human body owns, and it sits directly against tissue that is exquisitely sensitive to microbial intrusion. Life therefore depends on a boundary that is not a simple wall. It is a layered, active interface: mucus, antimicrobial chemicals, immune cells trained to tolerate the familiar and respond to the dangerous, and an epithelium that renews itself at remarkable speed. Health is not the absence of conflict; it is a managed truce.

Crohn's disease and ulcerative colitis, the two main forms of IBD, are often introduced through symptoms and scopes: abdominal pain, urgent diarrhea, blood in stool, weight loss, fatigue; inflamed tissue seen by colonoscopy; biopsies showing immune infiltration and damage. These details matter clinically, but they can obscure the deeper drama. IBD is what happens when the truce fails and the border stops functioning as a border. The immune system responds as it was designed to respond when microbes breach a barrier. The tragedy is that in the gut, microbes are always close. When the line between "nearby" and "inside" becomes fuzzy, the response can become chronic.

A historical thread helps explain how this way of thinking emerged. For much of the twentieth century, medicine was shaped by a powerful story: find the germ, kill the germ, cure the disease. Tuberculosis, cholera, diphtheria, syphilis—the germ theory revolution earned its dominance. When chronic intestinal inflammation was considered, it was natural to look for a single culprit bacterium. The failure to identify a consistent pathogen in ulcerative colitis and Crohn's disease was not for lack of trying. Patients could be desperately ill, their colons visibly inflamed, their blood chemistry screaming of systemic stress, and yet cultures did not yield an obvious invader that explained everything.

As molecular methods improved, the disappointment became data. Daniel Frank, Allison St Amand and colleagues (2007) used DNA-based approaches to show that IBD is associated with community-level imbalances in gut microbes, a phenomenon later bundled under the term *dysbiosis*—a shift away from a typical, diverse community toward a disturbed configuration. The crucial lesson was not that there is "an IBD microbiome" with a neat signature; there is not. Different patients show different patterns, and even the same patient can fluctuate over time. The lesson was that IBD looked less like an infection by a single organism and more like an ecosystem under chronic stress, with the host immune system and the microbial community locked in a destructive feedback loop.

The border that fails in IBD is often discussed in microscopic terms, and for good reason. The colon is lined by a single layer of epithelial cells, one cell thick, renewed every few days. This layer is not naked. In the large intestine, it is covered by mucus organized into two layers. The inner layer, close to the epithelium, is relatively dense and largely free of bacteria; the outer layer is looser and heavily colonized, providing a habitat that keeps microbes at arm's length. If the colon were a city, the mucus would be its buffer zone: not empty space, but structured territory that makes contact less likely and more controlled.

Mucus is made primarily of large glycoproteins called mucins. Its physical properties matter: thickness, viscosity, how fast it is renewed, how it is cross-linked. But mucus is also biochemical. It contains antimicrobial peptides and binds secretory IgA, an antibody class that coats microbes in a way that can limit their ability to approach the epithelium without necessarily killing them. The intestine also secretes defensins and other antimicrobial molecules, and it tunes local oxygen and nutrient gradients that help keep certain bacteria in check. The border is not passive; it is maintained.

When that maintenance fails, the consequences are immediate. Bacteria can come into close contact with epithelial cells. Molecules from bacteria—parts of cell walls, fragments of DNA, metabolites—

can be sensed by pattern-recognition receptors in epithelial and immune cells. These receptors are the immune system's burglar alarms. In a sterile tissue, their activation is appropriate and brief. In the gut, where bacterial signals are constant, activation must be carefully interpreted. A breach shifts interpretation from "background" to "danger." Neutrophils and other immune cells are recruited; inflammatory cytokines rise; tissue becomes swollen and fragile; bleeding becomes more likely; ulcers can form. In Crohn's disease, inflammation can be patchy and can involve deeper layers of the gut wall, sometimes leading to strictures and fistulas. In ulcerative colitis, inflammation is typically continuous and limited to the colon's inner lining, but can still be severe and life-altering. These patterns are medical facts; the ecological point is that both represent a breakdown of the normal compartmentalization between a microbial habitat and host tissue.

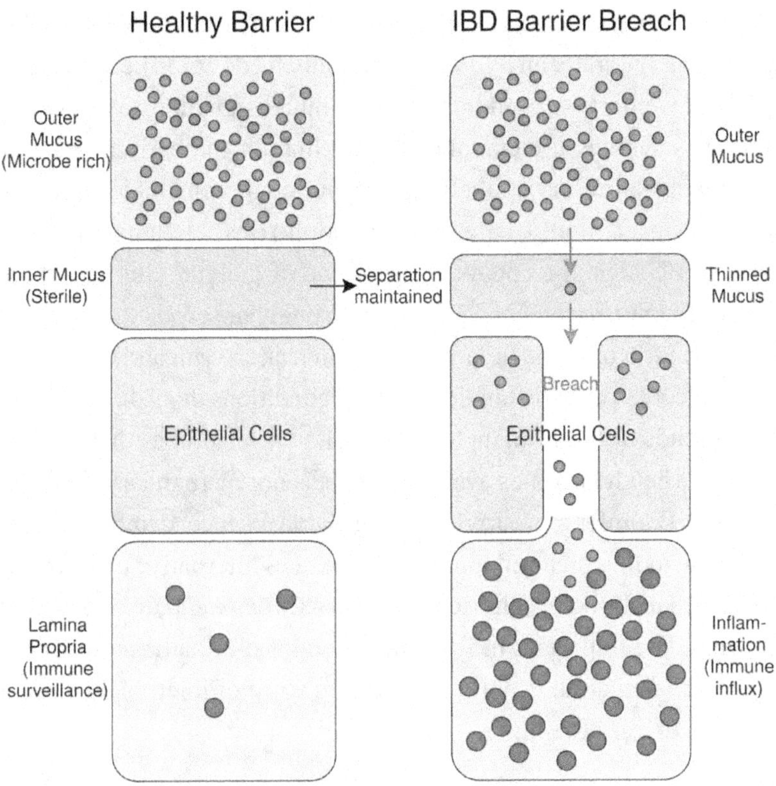

In real tissue the interface is more intricate than any diagram, but the principle is simple enough to hold in the mind. A healthy colon keeps most microbes in the outer mucus, away from epithelial cells. In IBD, the inner mucus layer can be thinned or disorganized, the epithelium becomes stressed, and bacteria and their products get too close for comfort. The immune system responds, and the response itself can make the environment even more hostile to the kinds of microbes that help maintain the barrier.

The phrase "keystone species" comes from ecology, where it describes an organism whose influence is disproportionate to its abundance. Remove it and the system may reorganize dramatically. The gut has its own candidates for keystone roles. Some microbes excel at producing butyrate, an SCFA that not only feeds colon cells but also tends to support anti-inflammatory immune signaling. When such microbes decline, a metabolic support line for the barrier weakens. The mucus layer and epithelium become more vulnerable; immune alarms trigger more easily; inflammation rises. Inflammation, in turn, can change the chemical landscape of the gut, altering which microbes are favored. A calm gut is strongly anaerobic, meaning it contains little oxygen. Inflammation can leak oxygen and other reactive molecules into the lumen, creating conditions that advantage different kinds of bacteria, including facultative anaerobes that can use oxygen when it becomes available. This is not mere theory. In 2010, Andreas Bäumler's group, including Sarah Winter, demonstrated a mechanism in which inflammation generates alternative electron acceptors that allow certain bacteria, such as *Salmonella*, to respire and expand. The important move is not the choice of pathogen; it is the logic: inflammation is not only a consequence of microbial shifts, it can be a driver of them.

This feedback loop explains why IBD can be so stubborn. When the gut is inflamed, it becomes a different habitat. The immune system releases antimicrobial molecules and alters nutrient availability. Blood and mucus can provide iron and other resources. Reactive oxygen and nitrogen species can act as new chemical opportunities for microbes equipped to exploit them. A community that thrives in that altered habitat may not be the community that best supports barrier integrity. What begins as a defensive response can end up selecting for organisms that perpetuate the need for defense.

That still leaves the question patients ask, often with a mix of anger and hope: why did it start? The honest answer is that there is no single beginning. Genetics contributes risk. Many IBD-associated genetic variants involve immune sensing, barrier function, and microbial handling. Environment contributes risk: diet, infections, stress, smoking (notably in Crohn's disease), medications, early-life microbial exposures. The microbiome contributes risk and modifies disease course. But these are not additive causes in a simple sum. They are interacting conditions that make a breach more likely and a breach more costly.

A concrete example shows the human stakes. A young adult develops intermittent abdominal cramps and diarrhea that becomes more frequent over months. They assume it is stress or a food intolerance, until one morning the stool is streaked with blood. A colonoscopy shows inflamed tissue; biopsies suggest ulcerative colitis. In the weeks that follow, daily life is reorganized around bathroom proximity, fatigue, and the fear of sudden urgency. Medications can help dramatically, but they also carry tradeoffs: steroids can calm an acute flare but have serious long-term side effects; immunomodulators and biologic therapies can reduce inflammation but increase susceptibility to certain infections. Behind the clinic visits is the border problem: a body trying, relentlessly, to keep microbes in their place.

Once IBD is framed as a border failure, two ideas that otherwise seem contradictory can coexist. On the one hand, reducing immune overreaction is essential; uncontrolled inflammation is damaging in its own right. On the other hand, supporting the barrier and the ecosystem that maintains it is also essential; otherwise immune suppression can feel like lowering the volume of an alarm while leaving the window open. This is why dietary strategies, while not cures, can matter in some patients: not because fiber or fat is morally good or bad, but because changing substrates changes fermentation, metabolites, mucus dynamics, and microbial composition. It is also why certain patients experience flares after gastrointestinal infections: an infection is a direct assault on the interface, and the ecosystem that reassembles afterward may not be the same as before.

There is a related idea with a deliberately provocative name: *metabolic endotoxemia.* In mouse models described by Patrice Cani and colleagues (2007), high-fat feeding was associated with a modest but chronic rise in circulating lipopolysaccharide (LPS), a component of the outer membrane of many Gram-negative bacteria. LPS is one of the immune system's classic danger signals. In those experiments, low-dose LPS infusion could mimic aspects of diet-induced metabolic disease, and genetic disruption of the CD14 pathway, involved in LPS sensing, blunted some effects. The details belong to metabolism as much as to IBD, but the conceptual bridge is useful: barrier function and immune sensing link the gut ecosystem to inflammation far beyond the intestine. The body is not surprised by bacteria in the colon; it is surprised by bacterial molecules in places they do not belong.

IBD can be understood as a more intense, localized version of the same principle: a persistent mismatch between where microbial products are and where the immune system expects them to be. When the interface is intact, immune education in the gut tends to favor tolerance. When the interface is leaky or inflamed, immune education tilts toward attack. Over time, the immune system can become primed to respond aggressively to microbes that were once treated as familiar neighbors.

It is tempting to ask whether there is a single villain microbe in IBD, and candidates have been proposed: adherent-invasive *E. coli* in some Crohn's disease patients, reductions in certain *Faecalibacterium* strains, blooms of opportunists during flares. The more robust view is that the villain is not an organism but a *state*: reduced diversity, loss of barrier-supporting functions, and an environment that selects for inflammation-tolerant microbes. In that state, even normally harmless microbes can become irritants simply by being in the wrong place at the wrong time, too close to epithelial cells that are already stressed and immune cells that are already on edge.

The most humane way to end is to return to the truce itself. The gut is not meant to be sterile. Sterility would be a kind of death. The task is coexistence under rules enforced by mucus, epithelium, and immune restraint. IBD is what it feels like when those rules collapse: the body treating its own inhabited interior as a wound that will not close, the microbes adapting to an increasingly hostile landscape, and the person caught between them, trying to live an ordinary day while an extraordinary argument rages at the edge of their cells.

7.3 The Opportunist

If the gut microbiome is an ecosystem, antibiotics are a wildfire. They are sometimes lifesaving, sometimes overused, and almost always ecologically blunt. A pill taken to treat pneumonia reaches the colon not as a polite, targeted emissary, but as a chemical disturbance that can thin entire populations, scramble food webs, and alter the chemistry of the habitat. Most of the time the forest regrows. Sometimes, though, the first thing to sprout is not a diverse canopy but a single aggressive weed.

Clostridioides difficile—often still called *Clostridium difficile*, a name many clinicians learned first—is the organism that made this ecological principle clinically unforgettable. It is not the most common bacterium in the human gut, and it is rarely a problem in a stable community. Yet in hospitals and nursing homes, after antibiotic courses, it can bloom with startling speed, turning the colon into an inflamed, toxin-soaked landscape. In its severe form it can cause pseudomembranous colitis, so named because the colon develops raised, yellow-white plaques: not a true membrane, but a grim mixture of inflammatory debris, fibrin, and dead cells that looks like one.

There is a reason *C. difficile* has become the emblem of microbiome collapse. It behaves like an ecological rule made flesh: remove diversity, remove competition, and an opportunist takes over.

The key concept is *colonization resistance*. The phrase sounds military, but it describes something ordinary: the way resident microbes make it hard for newcomers to establish themselves. In a healthy colon, niches are occupied. Nutrients are consumed quickly. Attachment sites on mucus are already used. Many microbes produce bacteriocins and acids that inhibit rivals. Others modify bile acids and other chemicals in ways that make the environment inhospitable to particular invaders. The immune system also participates, but often indirectly: a stable microbiome helps maintain the mucus barrier, tunes antimicrobial peptide production, and encourages a tolerant immune tone that is paradoxically better at handling true threats. Colonization resistance is not a single mechanism; it is the summed effect of a busy, crowded habitat.

C. difficile is adapted to wait out hard times. It forms spores, a dormant state that can survive heat, drying, and many disinfectants. Spores can persist on surfaces, travel on hands, and move through wards despite careful cleaning. A person may ingest spores and never know it, because the resident community and bile acid chemistry prevent the spores from germinating and the vegetative bacteria from thriving. After antibiotics, the chemical and ecological barriers shift. Suddenly there is space. Suddenly there is food. Suddenly the bile acid signals that once discouraged germination can change in ways that encourage it. The spore wakes up into a world with fewer competitors.

A concrete human story makes the stakes clear. An older patient is admitted with a urinary tract infection and receives a broad-spectrum antibiotic. The fever resolves. A week later, watery diarrhea begins, at first attributed to diet changes in the hospital or the stress of illness.

Within days the diarrhea becomes relentless; abdominal pain intensifies; blood pressure drops; white blood cell counts climb; creatinine rises as dehydration and systemic inflammation take hold. A stool test detects *C. difficile* toxin. Treatment begins, but recovery is not guaranteed, and recurrence is common. When people describe *C. difficile* infection, they often emphasize not just the physical misery but the psychological toll: the fear of leaving home, the embarrassment, the sense that one has been contaminated by the very place meant to heal.

The organism's punch comes from toxins, principally TcdA and TcdB, that disrupt the cytoskeleton of epithelial cells and trigger inflammation. The result is a colon that leaks fluid, sheds cells, and summons immune cells in huge numbers. The toxins turn an ecological bloom into tissue damage. This is one reason opportunists are so dangerous: once they seize the habitat, they can change it in ways that favor their continued dominance, at least long enough to spread.

The story has an important historical pivot. For decades, *C. difficile* was not widely recognized as the cause of antibiotic-associated colitis. Clinicians knew that antibiotics could provoke diarrhea, but the mechanism was unclear. In the late 1970s, work by several groups identified a toxin-producing organism in patients with pseudomembranous colitis, and *C. difficile* emerged as the culprit. The recognition reshaped hospital infection control and changed how clinicians thought about the collateral damage of antibiotics. An adverse effect was no longer merely a side effect in the patient; it was an ecological opening in the gut, with consequences for transmission in the ward.

If antibiotics create the opening, why does the community not simply snap back once the drug is stopped? Sometimes it does. Sometimes it does not, and the reasons are ecological. After a disturbance, the first colonizers can lock in the direction of recovery. If *C. difficile* establishes itself early, its toxins inflame the tissue, increasing oxygenation and changing nutrient availability in ways that can keep the community unstable. Meanwhile, the microbes that once provided resistance may be slow to return, either because they were wiped out, because their food sources are scarce, or because the new community configuration actively excludes them. Recurrence becomes a loop: antibiotics used to treat *C. difficile* can further suppress the recovering community, leaving the door open yet again.

A useful term here is *domination*. In a diverse microbiome, no single group should monopolize the habitat. After antibiotics, however, studies in mice showed that introduced organisms such as vancomycin-resistant *Enterococcus* (VRE) can sometimes expand until they occupy a startling fraction of the gut community. In high-risk human settings, such as patients undergoing intensive chemotherapy or allogeneic hematopoietic stem cell transplantation, similar "domination" patterns have been associated with subsequent bloodstream infections. The details vary by patient and organism, but the ecological warning is consistent: when diversity collapses, the gut can become a reservoir for microbes that do not stay politely in the lumen.

C. difficile is not always acquired from outside. Some people carry it at low levels without illness. Others acquire it from the environment. Either way, it thrives on the same opportunity: the absence of resistance. Thinking in terms of colonization resistance also clarifies why *C. difficile* risk is patterned socially and institutionally. It is higher where antibiotic exposure is frequent, where vulnerable people concentrate, where spores circulate, where proton pump inhibitors and other medications may alter gut conditions, and where nutrition and mobility are compromised. The microbe is opportunistic, but so is the setting.

Once the problem is framed as ecosystem collapse, the most effective treatment begins to look less like warfare and more like restoration.

This is where fecal microbiota transplantation (FMT) enters the story, with a plot twist worthy of a dinner party: a treatment that involves transferring stool from a healthy donor to a patient, with the goal of reintroducing missing community functions. The idea feels modern because it is associated with microbiome science, but it also has older roots. Historical reports from Chinese medicine describe the use of "yellow soup" for severe diarrhea centuries ago, a reminder that pragmatic medicine sometimes arrives before mechanistic understanding. In contemporary practice, FMT was propelled into mainstream attention by a landmark randomized trial in 2013 (van Nood and colleagues) showing striking efficacy of donor feces delivered via the upper gastrointestinal tract for recurrent *C. difficile* infection compared with standard antibiotic therapy. For many clinicians and patients, it was the moment the microbiome stopped being a research curiosity and became a therapeutic target.

The logic is straightforward. If colonization resistance is a community property, then a community may need to be replaced, not merely supplemented. Probiotics, in most cases, are a few strains. FMT is hundreds of species and their interacting metabolisms, transplanted as a functioning unit. It is a crude method by pharmaceutical standards, but it matches the problem: a system-level failure.

That does not mean FMT is simple or uniformly effective. Delivery matters: via colonoscopy, enema, nasoenteric tube, or capsules. Donor selection matters. Processing matters. Patient factors matter. More recent trials and real-world experience have shown variability in outcomes depending on the product and protocol. Safety is also central. Stool can transmit pathogens, including those that are hard to detect, and screening has become increasingly rigorous, especially after reports of transmission of multidrug-resistant organisms in immunocompromised recipients. The success of FMT has therefore accelerated a shift toward more standardized microbiota-based therapeutics: defined consortia of cultured bacteria, or purified spore preparations, designed to provide colonization resistance with less unpredictability.

Underneath these practical questions lies a mechanistic one: what exactly restores resistance? Part of the answer involves bile acids. In a healthy colon, resident microbes convert primary bile acids (made by the liver) into secondary bile acids. These transformations can influence *C. difficile* spore germination and vegetative growth. After antibiotics, the microbes responsible for these conversions can diminish, shifting the bile acid pool in a direction that favors *C. difficile*. Restore those microbes, and the bile acid chemistry changes back, closing an important gate. Other mechanisms likely contribute as well: nutrient competition (especially for simple carbon sources released after antibiotic-induced changes), production of inhibitory metabolites, and improved barrier function that reduces inflammation-driven habitat changes.

The *C. difficile* story also sharpens the meaning of the word *pathogen*. In classic infectious disease, a pathogen is an organism that causes disease when it enters the body. In the gut, context blurs the categories. Some organisms are obvious pathogens. Others are *pathobionts*: residents that can behave harmlessly in one setting and harmfully in another. Even *C. difficile* can be carried without symptoms, especially in infants, whose gut receptors and immune responses differ. The debate over what counts as normal is not academic. It determines how we think about risk, screening, and treatment. It also forces humility: what looks like a villain under one set of ecological conditions can be neutral under another.

Antibiotics will remain indispensable. The goal is not to romanticize microbes at the expense of human survival. The goal is to recognize that antibiotics act on a microbial organ, and organs have side effects when you injure them. This recognition changes clinical practice in small but meaningful ways. It encourages narrower-spectrum antibiotics when appropriate. It encourages shorter courses when evidence supports them. It motivates infection control that treats spores as ecological seeds. It prompts caution with repeated antibiotic exposure in vulnerable patients. And it makes room for therapies aimed at preserving or restoring colonization resistance rather than simply killing whatever is currently dominant.

There is a final emotional lesson in *C. difficile*: it makes visible the hidden labor of the microbiome. Most days, the gut community is a quiet public service, preventing blooms of opportunists without asking for thanks. People rarely feel colonization resistance; they notice it mainly when it is gone. Then the loss is not subtle. It is urgent, painful, and, in severe cases, lethal.

A healthy microbiome does not guarantee protection. A disrupted one does not guarantee disease. But *C. difficile* demonstrates that a complex community can function like a shield, and that shattering that shield can invite an organism whose greatest talent is patience—waiting as a spore for the moment the ecosystem forgets how to say *no*.

Chapter 8

Tending the Garden

We are not passive observers of our internal ecosystem, but its primary architects, wielding influence through every meal we consume. By understanding the distinction between transient microbial visitors and permanent residents, we can separate marketing hype from the biological realities of how to cultivate a resilient inner wilderness.

8.1 Feast or Famine for the Guilds

Imagine hosting a dinner party for a city you never get to meet. Every day, you slide plates through a mail slot. The guests remain behind a wall, but their mood determines whether the house stays peaceful or gets vandalized from the inside. That is the practical relationship most of us have with the gut microbiome: we choose the menu, and then we live with the consequences.

Diet is not merely a source of calories for *us*. It is the main set of environmental conditions for the microbes that live on and within the gut lining. If the intestines are a habitat, food is rainfall, soil chemistry, and the seasonal flowering of plants all at once. And as in any habitat, when the usual inputs disappear, residents improvise. The improvisation can be ingenious, even elegant, but it can also be costly for the host.

Microbes do not experience "nutrition" in the same way people do.

A nutrition label tells you about fat, carbohydrate, and protein in grams. A microbe cares about *structure*: whether a carbohydrate is a simple sugar it can grab immediately, or a tangled, branched polymer that requires specialized tools; whether a protein arrives packaged with plant cell walls or as a soft slurry from a burger; whether the day's menu offers a steady trickle of complex substrates or a burst of easily absorbed sugars that vanish in the small intestine before most microbes see them. The gut microbiome is shaped less by what *we* can digest than by what reaches the colon still intact.

That is why dietary fiber, for all its unglamorous reputation, sits near the center of the story. Many fibers are carbohydrates that human enzymes cannot break down. They pass through the stomach and small intestine like locked suitcases. Only when they arrive in the colon do microbes open them with enzymes humans never evolved. The technical term often used in research is *Microbiota-Accessible Carbohydrates* (MACs): carbohydrates that the microbiota can access and metabolize. The key idea is deceptively simple. If MACs are plentiful, microbes have a reason to specialize in fermenting them. If MACs are scarce, microbes that can live off other substrates—including host-derived ones—gain an advantage.

The microbiome is sometimes described as a list of species. In practice it behaves more like a set of *guilds*: groups of organisms that may not be close relatives, but that perform similar jobs in the ecosystem. In a forest, you can have many different pollinators; what matters is the function. In the colon, you can have many different fiber fermenters; what matters is the metabolic role they play, the products they make, and how those products affect the host. Guilds are a convenient way to think ecologically without pretending we know the full cast list in every individual.

One of the most consequential guilds is the community of fermenters that turn MACs into short-chain fatty acids (SCFAs). SCFAs are small molecules—mainly acetate, propionate, and butyrate—produced when microbes break down complex carbohydrates in oxygen-poor conditions. The colon is a fermentation vessel, and SCFAs are its signature output. Butyrate deserves special attention, not because it is a miracle molecule, but because it is a clear example of how microbial waste can become host treasure. Colon cells use butyrate as a major fuel. It also acts as a signal: it can influence immune tone and gene expression in ways that researchers are still mapping, including through receptors that sense these metabolites and through epigenetic mechanisms such as histone deacetylase inhibition. Translated into everyday language: butyrate is both *food* for the gut lining and *information* that the microbial community is well fed.

A well-fed fermentative guild tends to produce a chemical environment that discourages certain troublemakers. Fermentation lowers pH, making the colon more acidic. Many pathogens and opportunists do not enjoy that acidity. SCFAs and related metabolites can also help keep the gut barrier in good working order. The barrier is not a single thing; it is a layered system. There is a mucus layer, built from heavily glycosylated proteins called mucins, forming a slippery, protective matrix. Beneath that lies a single-cell-thick sheet of epithelial cells held together by tight junctions, the molecular rivets that decide what may pass between cells. Below that sits a vigilant immune presence, ready to respond when microbes get too close.

This is where the "feast" part of the story earns its name. When MACs are abundant, microbes compete to harvest them, and their success produces metabolites that help maintain the very habitat they depend on. It is a feedback loop that rewards cooperation at the ecosystem level without requiring any one microbe to be altruistic. Microbes do not need to "want" to help the host. They merely need to survive in a context where their favored food arrives reliably. The host, in turn, benefits from a metabolically active, well-contained community.

The "famine" state looks different. When MACs are chronically scarce—as can happen in dietary patterns that are low in diverse plant fibers—microbes do not vanish. They adapt. Some shift toward fermenting proteins and amino acids that arrive in the colon, producing a different metabolite profile, including compounds that at high levels can irritate the gut. But the more striking adaptation is that some microbes turn to *host-derived carbohydrates*. The gut, from a microbe's perspective, is not only a tube carrying food. It is a surface coated with glycans: sugar chains attached to mucins in the mucus layer. Those glycans are meant to be a buffer, a sacrificial interface that keeps bacteria at a safe distance from the epithelial cells. Yet they are also edible.

In animal experiments that helped sharpen this picture, fiber deprivation shifted microbial behavior toward mucus consumption. The mucus layer became thinner and more penetrable. Once that protective layer is compromised, bacteria and their molecules can come closer to the epithelial surface, increasing the likelihood that the immune system interprets the situation as a breach. The result can be a low-grade inflammatory tone and increased vulnerability to pathogens that exploit access to the gut wall. This is not the same as claiming that a low-fiber day "causes" disease. It is a description of ecological pressure: remove one set of food resources and you select for microbes that can use another.

The concept is intuitively unsettling because it reframes diet as a kind of domestic policy. When you do not feed the guilds that thrive on MACs, you are not simply "missing out" on benefits. You are changing who prospers in the neighborhood, and what the neighborhood must consume to keep going. The microbiome is remarkably resilient, but resilience has limits. Ecologists have long known that repeated disturbances can push an ecosystem into a different stable state, one that does not bounce back easily even when conditions improve. In the gut, this can translate into reduced diversity and loss of certain fiber-degrading capacities, especially if low-MAC patterns persist across long periods.

One reason diet exerts such a strong selective pressure is speed. In a famous demonstration of how quickly microbial communities respond, carefully monitored dietary shifts altered measurable aspects of the human gut microbiome within days. That finding, often surprising to the public, aligns with what any microbiologist expects: microbes have short generation times and fierce competition. Give them a new substrate, and the organisms best equipped to exploit it can bloom rapidly. Remove that substrate, and the bloom collapses just as fast. From the host's perspective, the microbiome can feel like a stubborn personality trait. From the microbe's perspective, it is a market responding to prices.

A concrete modern example is the contrast between a fiber-rich meal that delivers complex carbohydrates to the colon and a highly processed meal where most carbohydrates are refined starches and sugars absorbed upstream. Consider the difference between a bowl of intact oats with berries and nuts versus a sugary drink and a pastry. Both can contain "carbs," but only one reliably sends a substantial portion of complex material to the colon. The colon's residents do not care about the glycemic index; they care whether anything arrives for them at all. When the colon receives little fermentable substrate, it becomes metabolically quieter in one sense (less SCFA production), but more opportunistic in another (more pressure to use alternative fuels).

It is tempting to talk about "good bacteria" and "bad bacteria" here, but that framing misleads. Many microbes have dual potential. A mucin-degrader is not necessarily a villain; in a fiber-rich environment, controlled mucin turnover is part of normal biology, and certain mucus specialists may help keep the mucus layer renewed. The trouble is not the existence of these organisms; it is the *context* that makes them dominant or overactive. Ecology is the art of context.

There is also a historical angle that illuminates why modern diets can create an ecological mismatch. For most of human history, diets contained a wide variety of plant materials that were minimally processed: roots, leaves, seeds, tubers, and seasonal fruits, often accompanied by the inevitable grit and complex polysaccharides that come with eating food close to its original form. The industrialization of food changed the physical form of carbohydrates. Milling, refining, and extrusion made calories cheap and consistent, but they also shifted where digestion happens. When flour becomes ultra-fine and sugars become ubiquitous, more of the carbohydrate load is absorbed early, leaving less work for the colon's fermenters. It is a subtle shift in anatomy's division of labor: the small intestine does more, the colon does less, and the microbial guilds adapted to MAC abundance find themselves underemployed.

The public health consequences of industrial food are not reducible to the microbiome, and it would be irresponsible to claim otherwise. But the microbiome provides a mechanistic bridge between dietary pattern and physiological state. If you feed fermenters, you tend to increase SCFAs and support barrier function. If you starve them, you tend to reduce SCFAs and increase reliance on host-derived substrates. Those are not moral judgments. They are predictable outcomes of resource availability.

A useful way to think about MACs is as the microbiome's version of infrastructure spending. You are not paying individual workers directly; you are funding the roads, power lines, and water systems that allow certain industries to exist. Different fibers favor different industries. Some microbes specialize in breaking down inulin-type fructans; others are better at resistant starch; others thrive on pectins and arabinoxylans. Often, no single organism completes the job alone. One microbe starts by chopping a large polymer into smaller pieces; another consumes the fragments; a third converts intermediate products into butyrate. This cross-feeding is a hallmark of microbial communities. The gut is not a solitary workshop but a supply chain.

That supply chain has vulnerability points. If the early degraders disappear, downstream butyrate producers may struggle even if the fiber is present, because they depend on intermediate products. This helps explain why two people can eat similar high-fiber diets and yet show different metabolic outcomes in their stool metabolites. The ecology is personal, shaped by history: antibiotics, early-life exposures, infections, and long-term diet all influence which guild members are available to take up a job when the substrate appears.

The mucus layer deserves one more moment in the spotlight, because it is easy to treat it as mere slime. In fact it is a highly engineered interface, continuously produced and remodeled. It acts like a biological "no man's land," limiting bacterial contact while still allowing controlled chemical communication. When microbes erode it under fiber scarcity, the consequence is not simply thinning. The host may respond by altering mucus production, immune signaling, and epithelial turnover. That response can be protective, but it can also create a more inflamed environment that favors different microbial strategies, reinforcing dysbiosis. Again, feedback loops appear: diet changes microbial behavior, microbial behavior changes barrier status, barrier status changes immune tone, immune tone changes microbial selection.

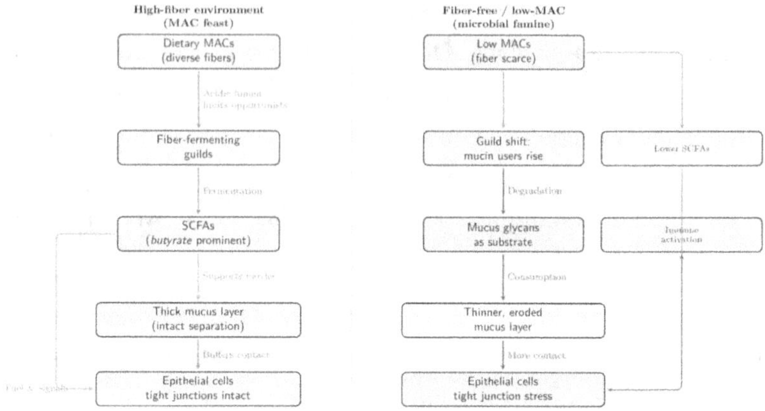

The picture is necessarily simplified, but it captures a crucial asymmetry. Humans can tolerate short periods of dietary monotony. The microbiome can too. The risk lies in making famine the default condition for guilds that sustain a calm, well-fortified interface. When that happens, the ecosystem does not merely idle; it reorganizes around whatever resources remain dependable.

For a tangible example, think of what happens during a typical modern "reset" attempt. Someone eats a very low-carbohydrate diet with little fermentable fiber for weeks, then swings back to a weekend of high plant intake. The human body may experience this as a welcome return of bowel regularity. The microbes experience it as a shock: a sudden flood of substrates after scarcity. Some fermenters bloom, gas production spikes, and the person concludes that fiber "doesn't agree" with them. Yet the discomfort may reflect not that fiber is inherently harmful, but that the microbial infrastructure for processing it has been underfunded. A steadier reintroduction often proves more tolerable because it allows guilds to expand without explosive fermentation.

There is an older, almost accidental historical anecdote that illustrates how much these invisible processes matter. In the nineteenth century, physicians argued fiercely about "autointoxication," the idea that the colon produced poisons that seeped into the body, causing everything from fatigue to depression. Much of that theory was wrong, and some of its treatments were horrific. Yet buried in the mistaken framework was a correct intuition: the colon is a chemical factory, and the products of fermentation and putrefaction (microbial breakdown of proteins) can influence physiology. Today we would reject the melodrama of self-poisoning, but we can acknowledge the kernel of truth: when microbial metabolism shifts, the host experiences a different internal chemistry. Modern tools finally allow us to measure these molecules rather than infer them from vague symptoms.

The most practical implication is also the least glamorous: consistency matters. A microbiome fed with a variety of MACs tends to maintain guilds that produce SCFAs and support barrier function. The variety matters because different fibers select for different enzymatic toolkits. A single supplement cannot replicate the structural diversity of plants. Even within one category—say, "whole grains"—the difference between intact kernels, rolled grains, and ultrafine flour alters what reaches the colon. Food processing changes microbial access as surely as it changes taste.

None of this requires turning meals into a laboratory experiment, or treating fiber as a magic charm. The point is ecological. The gut microbiome is not impressed by intentions; it responds to substrates. Feed the guilds that thrive on complex carbohydrates, and the ecosystem tends to generate metabolites that keep the interface between microbe and human stable and quiet. Starve them, and the same ecosystem will keep itself alive by consuming what it can reach—including, at times, the host's own protective layers. The difference between feast and famine is written less in calories than in the molecular shapes that arrive at the far end of the gut, and in the microbial jobs those shapes make possible.

8.2 Tourists, Not Residents

A probiotic is often sold with the language of restoration: a few capsules, a few billion “good bacteria,” and the gut will be repopulated, as if an empty garden bed were being reseeded. The metaphor is comforting because it suggests a straightforward fix. Swallow the right organisms and they will move in, pay rent, and keep order.

The trouble is that most adult guts are not empty lots. They are densely inhabited ecosystems with rules, routines, and bouncers. When new microbes arrive, even ones with an excellent reputation, they meet a community that has spent years defending its territory. The usual outcome is not permanent colonization but transient passage: a brief visit, a measurable presence in stool for a day or two, and then disappearance once the capsules stop. Calling this a failure can miss the point. A tourist can still spend money, cause a traffic jam, and trigger a police response. Transient microbes can still send signals to the immune system, alter metabolites while they are present, and sometimes help in narrowly defined clinical contexts. They just rarely become long-term residents.

It helps to start with a simple question that advertisements glide past: what does it mean for a bacterium to “work” in the gut? There are at least two distinct possibilities. One is residency: the organism establishes itself, reproduces, and persists as part of the person’s baseline community. The other is a temporary effect: the organism passes through but interacts with the host and the resident microbes along the way, shifting physiology for as long as it is present. Many commercial narratives quietly promise the first while the evidence more often supports the second.

The adult colon is crowded beyond intuitive imagination. Microbes occupy the lumen (the interior space), but the most valuable real estate is near the mucus and epithelial surface where nutrients, oxygen traces, and host signals differ from the free-flowing contents. That surface-adjacent zone is also where the immune system pays closest attention. To become a resident, a probiotic strain has to survive stomach acid and bile, reach the colon alive, compete with established organisms for food, avoid being flushed out by peristalsis, find a niche that is not already occupied, and persist despite immune surveillance. That is a tall order for any outsider.

Ecologists call the phenomenon that makes invasion difficult *colonization resistance*. In everyday terms: an established ecosystem resists newcomers. In the gut, colonization resistance is achieved through several overlapping mechanisms. Residents consume available resources quickly, leaving little for an invader. They occupy attachment sites, making it physically hard for outsiders to get close to the mucosa. They modify the chemical environment, for example by keeping pH low through fermentation. They also produce antimicrobial compounds and stimulate the host to do the same. The result is not a single gatekeeper but a layered defense system that makes long-term engraftment uncommon.

A paradox follows from this. Many people who are told they have a "bad microbiome" likely have a microbiome that is, from an ecological standpoint, quite robust: it resists invasion. Robustness is not the same as health, and resistance is not the same as ideal composition, but the point stands. The gut community is often stable enough that a newcomer cannot simply move in because a label says it is beneficial.

This mismatch between marketing and ecology has a history. At the beginning of the twentieth century, the Russian-born scientist Élie

Metchnikoff, working at the Pasteur Institute, became fascinated by the idea that certain microbes might promote longevity. He noted the consumption of fermented milk in some rural populations and speculated that lactic acid bacteria could suppress harmful intestinal putrefaction. Metchnikoff's ideas were not modern microbiome science, but they were strikingly prescient in one respect: they framed gut microbes as modifiable partners in health rather than mere contaminants. They also helped popularize the notion that swallowing live bacteria could change the gut for the better. A century later, the cultural echo remains. Yogurt cups and supplement bottles often carry a faint, unspoken promise of Metchnikoff's optimism.

What has changed is not the human appetite for a simple intervention, but the scientific ability to test what actually happens to introduced strains. Modern studies can track strains through DNA sequencing and can even sample the gut mucosa directly, rather than relying solely on stool. That last distinction matters because stool is not the gut wall. Stool is what has been expelled. It is a useful window, but not a full map.

When researchers have done invasive, carefully controlled studies in humans, a consistent picture emerges. Probiotic strains can survive passage through the gastrointestinal tract and can be detected in stool during ingestion. Yet mucosal colonization—the kind that would suggest true residency—is highly variable from person to person and often limited or absent. Even when the same strains are given to many individuals, some people show short-term engraftment in certain regions of the gut while others show almost none. The effects, when present, can also be transient, fading after the probiotic is stopped. That variability is not a nuisance detail; it is the result. It tells you that the host and the resident ecosystem set the terms of entry.

This is where the word "probiotic" deserves careful handling. Scientifically, a probiotic is defined as a live microorganism that, when administered in adequate amounts, confers a health benefit on the host. That definition is narrower than the marketplace. It does not say the organism must colonize, and it does not say every product with live bacteria qualifies. It insists on a demonstrated benefit in a specific context. Without that, "probiotic" becomes a mood rather than a claim.

The colonization question is not just academic. Many people take probiotics after antibiotics, assuming they will help restore what antibiotics disrupted. The logic sounds reasonable. Antibiotics are a wildfire; probiotics are new seedlings. But ecology again complicates the story. After antibiotics, the gut community is altered, and colonization resistance may be temporarily weakened. Outsiders may find it easier to gain a foothold. That sounds like an opportunity, but it also raises a risk: if the foothold belongs to the wrong organisms, recovery could be distorted.

A striking example comes from work by Eran Elinav's group, including studies led by Suez and Zmora (2018), which examined what happens after antibiotics when people take multi-strain probiotics. In that study design, the researchers did something rare: they looked not only at stool but also at the gut mucosa, and they compared probiotic use to spontaneous recovery and to autologous fecal microbiota transplantation (returning a person's own pre-antibiotic stool sample). The results challenged the idea that probiotics are a universal restorative. Some individuals showed delayed recovery of their indigenous microbiome and mucosal gene expression when taking probiotics compared with those who simply waited. Autologous FMT, in contrast, could accelerate aspects of recovery. The lesson is not that probiotics are "bad." The lesson is that in a disturbed ecosystem, adding organisms is not automatically beneficial and can sometimes interfere with the natural trajectory of regrowth.

This kind of finding is uncomfortable because it collides with an everyday experience: many people report feeling better on probiotics. The discrepancy can be real without implying that one side is dishonest. Symptoms such as bloating, stool frequency, and abdominal discomfort are influenced by many factors: placebo effects, changes in diet that accompany supplement use, regression to the mean, and genuine physiological shifts caused by transient microbes. A probiotic can reduce diarrhea for some people in some contexts, even if it does not permanently colonize. A transient visitor can still produce lactic acid, consume certain sugars, or interact with immune receptors while passing through. But the more sweeping claim—that probiotics "reseed" a depleted microbiome in a durable way—is harder to support.

To understand how a tourist can have an effect, consider what the immune system "sees" in the gut. The gut-associated immune system is not waiting for microbes to disappear; it is waiting for them to behave. It continually samples microbial molecules through pattern-recognition receptors that detect conserved features such as lipopolysaccharide, flagellin, and certain cell wall components. This surveillance does not necessarily lead to inflammation. Much of it leads to tolerance, training, and calibrated readiness. When a new strain arrives in high numbers, it can change the pattern of molecules and metabolites present at the mucosal surface. That can shift immune signaling, even if the strain is gone a week later.

There is also the matter of microbial conversations. Residents do not merely compete; they communicate chemically, sometimes antagonistically. A probiotic strain might secrete bacteriocins (small antimicrobial peptides) that inhibit competitors while it is present. Or it might consume a substrate that otherwise would fuel gas-producing organisms, reducing bloating. Or it might produce metabolites that influence motility. These are plausible mechanisms for short-term symptomatic changes that do not require long-term residence.

But these mechanisms also underscore why effects can be inconsistent. A visitor that consumes lactose will matter in someone who eats lactose and lacks lactase, but do little in someone who does not eat dairy. A visitor that produces a certain antimicrobial will matter only if the targeted organisms are present and sensitive. A visitor that stimulates a particular immune pathway will matter only in a host whose immune baseline makes that pathway relevant. In other words, probiotics interact with a moving target: the existing ecosystem and the host.

A concrete example helps. Take *Lactobacillus rhamnosus* GG, one of the best-studied probiotic strains. It has evidence for benefit in certain settings, such as reducing the risk of some forms of antibiotic-associated diarrhea, especially in children, though results vary by study and population. When it is ingested, it can be recovered in stool during consumption and shortly after, which reassures consumers that it "survived." Yet persistence months later is not typical. If someone experiences benefit, it is usually because the strain is doing something while it is present: altering fermentation patterns, interacting with the mucosa, modulating immune responses, or competing with opportunists during a vulnerable window. The clinical claim, when justified, is often about reducing risk or shortening duration, not rebuilding an ecosystem from scratch.

This perspective also clarifies why dosing language in probiotic marketing can be misleading. "Ten billion CFU" sounds like a lot. It is. But the colon contains on the order of trillions of microbes. Even large doses may be numerically modest relative to the resident population, and numbers alone say nothing about ecological fit. Invasion is not won by headcount; it is won by access to resources and niches.

The obsession with CFUs also distracts from a more meaningful question: what is the strain's *job description*? Does it metabolize a substrate that will actually reach the colon? Does it produce metabolites likely to matter in that person's gut environment? Does it carry traits that help it adhere to mucus, or is it mainly a free-floating organism that will wash through? Is there evidence for benefit in the specific condition being targeted, at the dose and formulation being used? These questions are harder to put on a label, but they are closer to biology.

There is a second, often overlooked point: some of what people hope to achieve with probiotics is more plausibly achieved by feeding existing residents. That is where *prebiotics* come in. A prebiotic, as defined in an influential consensus led by Gibson and Hutkins (2017), is a substrate selectively utilized by host microorganisms that confers a health benefit. The word "selectively" matters. Not every fiber is prebiotic, and not every supplement that contains fiber-like ingredients will have a demonstrated benefit. Still, the conceptual contrast is useful. Probiotics attempt to add organisms. Prebiotics attempt to favor certain guilds already present. If colonization resistance makes seeding difficult, fertilizing may be a more reliable lever.

Yet even prebiotics can expose the complexities of the ecosystem. A person may take inulin and experience bloating. That does not automatically mean inulin is harmful; it may mean that certain fermenters rapidly produce gas in that individual's configuration of guilds. Another person may feel no discomfort and may show increased butyrate production. These differences are not failures of the concept but reminders that ecosystems are individualized. The same input can yield different outputs depending on who is already there and what cross-feeding networks exist.

Probiotics, in this light, begin to look less like permanent immigrants and more like seasonal workers. They can be useful, sometimes impressively so, in specific settings. There is reasonable evidence for particular strains in certain forms of infectious diarrhea, for prevention of necrotizing enterocolitis in preterm infants in some neonatal units (with careful product selection and safety oversight), and for certain symptoms in subsets of people with irritable bowel syndrome. But the evidence is not uniform, and the intervention is not one-size-fits-all. The strongest claims in the field tend to be narrow, strain-specific, and condition-specific. Broad claims tend to be broad precisely because they are harder to falsify.

There is also a safety dimension that tends to be minimized in consumer discussions. For most healthy people, probiotics are low risk, but "low" is not "zero." In immunocompromised individuals, those with central lines, critically ill patients, or very premature infants, live microbes can occasionally translocate and cause bloodstream infections. The rare nature of these events makes them easy to dismiss, but rarity is not reassurance when the product is taken casually by people whose underlying risk is unknown or changing. A microbiome intervention is not automatically benign because it is sold over the counter.

So why do the tourist microbes often fail to settle? Part of the answer is simply that humans and their microbes co-evolved under conditions that favored stability. A stable microbiome protects the host against pathogens by filling niches, consuming resources, and maintaining a hostile environment for invaders. From the host's viewpoint, colonization resistance is a feature. It means that every mouthful of food, every kiss, every swimming pool, does not rewrite the gut's ecology. If the gut were easily colonized by whatever passed through, we would be in constant biological flux, and not in a good way.

Another part of the answer is that commercial probiotic strains are often chosen for reasons that have little to do with colon ecology. Many are easy to culture, survive manufacturing, and tolerate oxygen, which helps them sit on shelves. But the organisms that dominate the adult colon are frequently strict anaerobes that dislike oxygen and are hard to package. The marketplace therefore favors strains that are convenient rather than representative. That does not make them useless, but it explains why they might act as visitors rather than settlers.

The most honest way to talk about probiotics is to treat them as biological signals delivered in living form. Sometimes a signal is all you need: a temporary push toward a different immune tone, a brief competitive check on an opportunist, a short-lived change in metabolite production. In other cases, especially when the goal is to restore a damaged ecosystem, signals may not be enough. The gut is not a flowerpot where you can replace the soil with a scoop of whatever you bought last week. It is a living organ-like ecosystem with memory.

At the dinner party, the guests behind the wall are already seated. Sending a few charming newcomers through the door does not guarantee they will be offered a chair. More often they stand near the entrance, exchange a few words, change the atmosphere slightly, and then leave. If the timing is right—after a disturbance, during an infection, in a particular host—that brief encounter can matter. If the timing is wrong, the encounter is simply a visit, and the city behind the wall goes on running itself, as it usually does.

8.3 The Ecological Reset Button

A gardener can usually coax a struggling plot back to life with small, patient changes: better watering, richer compost, fewer shocks. That logic works surprisingly well for the gut when the ecosystem is merely out of balance. Feed the microbes that make calming metabolites; stop starving the ones that keep the mucus barrier thick; avoid unnecessary antibiotic wildfires. Yet there are moments when the analogy breaks down, because the problem is no longer a struggling garden but a garden that has been paved over. In those moments, adding a few friendly species is like scattering seeds on concrete.

Fecal Microbiota Transplantation (FMT) sits at that uncomfortable boundary between gentle tending and radical reconstruction. It is often described in headlines as a kind of microbial "reboot." The word has drama, but it also misleads. Reboot implies returning to a known default. In reality, FMT is closer to a tissue transplant: replacing much of the microbial organ's living substance with a new community that can do the jobs the damaged one no longer performs. The shock value of the method sometimes distracts from how sober the medical rationale is. When it works, it works because the gut's ecology has rules, and FMT exploits them.

The cleanest illustration comes from *Clostridioides difficile* infection, usually abbreviated to *C. diff*. *C. diff* is not exotic. It is an opportunist that thrives when the normal community is disrupted, most famously after antibiotics. The bacterium forms hardy spores that can persist in healthcare environments and can survive the chemical and physical abuses that kill many other microbes. Once established, it can produce toxins that inflame and damage the colon, causing diarrhea that ranges from miserable to life-threatening.

Recurrent *C. diff* is where ecology stops being a metaphor and becomes a clinical fact. Many patients improve with standard antibiotic treatment directed at *C. diff*, only to relapse weeks later. The cycle can repeat, each round narrowing life to proximity to a bathroom, and for older adults in particular it can mean dehydration, hospitalization, and cascading loss of independence. The pain is not only physical. People describe fear of leaving the house, fear of eating, fear of being too far from help. If a disease can shrink a life, recurrent *C. diff* does it efficiently.

Why is it so hard to escape? It helps to imagine two different "resting states" of the gut community: a state that resists *C. diff* by default, and a state that leaves gaps *C. diff* can exploit. Antibiotics can push the gut from the first into the second. Once the second state is established, small nudges may not be enough to climb back over the barrier, especially if the person is repeatedly exposed to antibiotics, hospitalized, or otherwise stressed. This is an ecological idea with a formal name: *alternative stable states.* The details in the gut are more complex than any diagram, but the overall pattern is familiar to ecologists studying lakes, reefs, and forests. Some changes are reversible with gentle pressure; others require a shove.

Historically, medicine arrived at this insight in a roundabout way. One of the earliest modern case series is often traced to 1958, when Ben Eiseman and colleagues reported using fecal enemas to treat patients with what was then called pseudomembranous enterocolitis. The report predated today's understanding of *C. diff* toxins and spores, and it appeared in a medical culture that did not talk about "microbiomes." Yet the practical observation was hard to ignore: introducing fecal material from a healthy donor could rescue patients who were desperately ill. The intervention looked crude, even outrageous, but the result hinted that the problem was not a single pathogen in isolation. The problem was an ecosystem that had lost its normal capacity to defend itself.

With today's lens, the logic becomes clearer. A healthy gut community does not merely occupy space; it actively suppresses invaders. It competes for nutrients, transforms bile acids, acidifies the environment through fermentation products, and stimulates the host's barrier defenses. *C. diff* does not simply "arrive" and take over; it exploits a landscape made permissive by disruption. FMT works when it reintroduces a community capable of closing the gaps.

This is a crucial shift in mindset. The usual pharmacological approach is to identify a culprit and kill it. In recurrent *C. diff*, the more durable solution is often to restore the ecological context that makes *C. diff* an afterthought. That does not mean antibiotics are irrelevant; they are often necessary to knock down the immediate burden of *C. diff* before a restorative community can take hold. But it does mean the long-term therapeutic target is not only the pathogen. It is the microbial organ that normally keeps pathogens in check.

The distinction also clarifies why standard consumer strategies feel inadequate in this setting. Probiotics, in many people, behave like tourists: transient, inconsistent colonizers in a crowded landscape. Diet can shift guild behavior, sometimes quickly, but diet alone cannot always repopulate missing functions if key organisms have been lost or if the patient is too sick to eat and absorb normally. In the face of a collapsed ecosystem, adding a strain or two is like hiring a single security guard when the entire city's infrastructure has failed.

FMT, by contrast, imports infrastructure.

A good way to understand the force of that import is to think about "colonization resistance," the gut's tendency to repel newcomers. In a healthy adult, colonization resistance makes it hard for a probiotic strain to settle. In recurrent *C. diff*, colonization resistance has been weakened or misdirected. The ecosystem has lost some of its ability to say "no." FMT exploits that vulnerability in a controlled way: it introduces a large, diverse consortium of organisms that can rapidly occupy niches, consume resources, and re-establish the chemical environment that disadvantages *C. diff*. The same ecological gatekeeping that blocks most probiotics becomes an ally once a healthier community is installed, because the new community can then resist *C. diff* and other invaders.

None of this is magic; it is simply scale. A multi-strain probiotic might contain a handful of species chosen for manufacturability. A fecal transplant contains a complex living community, including organisms that are difficult to culture and that may be far more representative of the adult colon's true residents. It also includes bacteriophages and other microbial components that may matter for stability, though the science of those contributors is still maturing. In ecological terms, FMT is not "adding a species." It is attempting to replace a functioning network.

That ambition raises an obvious question: if FMT is so powerful, why not use it for everything? The short answer is that power is not the same as appropriateness. FMT makes the most sense when there is strong evidence that an altered community is a key driver of disease, and when the alternative treatments have failed or are inferior. Recurrent *C. diff* meets that standard unusually well. Many other conditions associated with microbiome differences—inflammatory bowel disease, irritable bowel syndrome, obesity, autoimmune disorders—do not. In those cases, microbiome changes may be consequences as much as causes, or only one thread in a tangled physiology. A dramatic ecological intervention is harder to justify when the causal chain is uncertain.

Medical societies have tried to reflect that restraint. The American Gastroenterological Association's clinical practice guideline on fecal microbiota-based therapies, published February 21, 2024, supports use for immunocompetent adults with recurrent *C. diff* after completing standard-of-care antibiotics, while discouraging routine use for conditions like IBD or IBS outside clinical trials. The dates matter because they mark a transition: FMT moved from improvised rescue to a more formal therapeutic category, with clearer boundaries around when it is justified.

Regulatory history reinforces the same point. In the United States, the Food and Drug Administration approved REBYOTA (fecal microbiota, live-jslm) on November 30, 2022, for prevention of recurrent *C. diff* infection in adults following antibiotic treatment for recurrent *C. diff*. Then, on April 26, 2023, the FDA approved VOWST (fecal microbiota spores, live-brpk), an oral product also indicated to prevent recurrence after antibacterial treatment. These approvals did not merely add convenience. They signaled that microbiota-based therapies could be standardized, manufactured, and evaluated more like conventional medical products, rather than handled as bespoke procedures dependent on local donor pools.

That standardization matters because the risks of microbiota transfer are real. A transplant is not only what you intend to give; it is what comes along for the ride. The FDA has issued safety communications over the years describing transmission of serious infections through investigational FMT, including multidrug-resistant organisms and pathogenic *E. coli* strains, with at least one reported death in an immunocompromised recipient. These events are rare, but they are the kind of rarity that should shape policy: devastating when they occur, and preventable only through rigorous screening, testing, and careful patient selection.

The gut's intimacy with the immune system makes this more than a contamination problem. Introducing a new community can provoke immune responses even when no overt pathogen is transmitted. For some patients, that immune perturbation may be neutral or beneficial; for others, it may be unpredictable. The immune system is trained on the person's existing microbial patterns. Replacing those patterns changes the daily antigenic weather.

The practical consequence is that FMT is not a wellness hack. It is a medical procedure (or a medical product) with a strong track record in a specific, high-stakes condition, and a less certain role elsewhere. Thinking of it as an ecological reset button tempts people to press it for vague symptoms, but the metaphor is safer if it keeps its full meaning: a reset is something you do when the system is locked in a malfunctioning state and the usual controls cannot recover it.

Even within *C. diff*, the picture is more nuanced than a miracle cure. Some patients relapse even after FMT. Some require repeat treatments. Some have co-morbidities that complicate the choice. And even successful FMT does not necessarily restore the gut to the patient's original baseline; it may create a new stable state, often healthier, but not identical. That is not a failure. It is simply how ecosystems work. When a forest regrows after a fire, it may not have the same tree composition as before. Function can return even when the species roster changes.

One of the most fascinating developments is the movement away from stool as the starting material. Stool is biologically variable and culturally fraught. It is also, from a mechanistic perspective, a black box. If the goal is to restore functions—bile acid metabolism that suppresses *C. diff*, production of key metabolites, occupation of niches—then, in principle, one could design defined consortia: carefully selected microbes assembled into a reproducible community. That idea has been pursued for years in research, and the FDA-approved products for recurrent *C. diff* represent a step toward a more controlled future. A capsule of spores is not identical to donor stool, but it embodies the same ecological strategy: restore a community capable of resisting *C. diff*.

A final, humbling point is worth keeping close. FMT's success in recurrent *C. diff* does not mean we can casually engineer the microbiome for any desired outcome. It means there exist situations where the microbiome's influence is strong enough, and the dysfunction discrete enough, that replacing the community restores a lost function. That is both inspiring and limiting. Inspiring, because it demonstrates that the microbiome is not merely correlated with disease but can be a lever for cure. Limiting, because it reminds us that most of the time the microbiome is woven into physiology so deeply that swapping it is not like replacing a battery; it is like transplanting part of an organ.

For someone trapped in the exhausting loop of recurrent *C. diff*, that distinction can feel abstract until the day symptoms finally stop returning. The relief is often described in practical terms: being able to drink coffee again, to travel without fear, to sleep through the night, to stop negotiating with the nearest bathroom. A functioning microbial community is easy to ignore when it is doing its job. When it collapses, the absence is loud. Restoring it can be one of the quietest triumphs in modern medicine: not the drama of a new drug that annihilates an enemy, but the steadier peace that comes when an ecosystem resumes its ordinary work of keeping the host alive and untroubled.

Chapter 9

Beyond the Hype

Microbiome science often dominates headlines with promises of panaceas, yet the path from a petri dish to a prescription is fraught with complexity. We examine the sophisticated detective work required to distinguish a true health driver from a harmless passenger, revealing the rigorous experiments that separate facts from fads.

9.1 Reading the Genetic Census

Imagine trying to understand a bustling city by peering through a keyhole. You might catch the sound of traffic, the rhythm of footsteps, a burst of laughter. But you would miss the street map, the power grid, the supply chains, the quiet conversations in cafés that determine tomorrow's mood. For most of the history of medicine, the microbial life inside the gut has been approached through a similar keyhole: grow what you can on a plate, count the colonies, and infer the rest.

That approach worked brilliantly for some of the most famous pathogens. *Vibrio cholerae* and *Mycobacterium tuberculosis* have strong opinions about where they like to live, and with enough persistence a microbiologist can often meet them on their terms. Yet the gut is not a sparse desert punctuated by a few villains. It is closer to a densely populated ecosystem with countless species, many of them fussy about oxygen, temperature, pH, neighbors, and nutrients. A surprising proportion will not grow in standard laboratory conditions at all. Others can be coaxed into growth only if you guess exactly what they eat, which gases they tolerate, and which companion microbes they depend on. Even when culture succeeds, it can flatter the organisms that thrive on the laboratory menu while ignoring the ones that do essential work in the shadows.

The great shift in microbiome science came from a simple, radical decision: stop asking microbes to introduce themselves by growing. Instead, take what they leave behind—their genetic material—and read it directly. This is why so much of modern microbiome research looks, at first glance, like it belongs in a genomics lab rather than a clinic: tubes of extracted DNA, machines that turn molecules into digital sequences, and algorithms that translate strings of A, C, G, and T into an inventory of life.

This inventory is often described with a bureaucratic phrase: the *genetic census*. Census-taking sounds dull, but the leap it enabled was as dramatic as switching from eyewitness rumor to satellite imagery. It made the invisible visible, not because it showed us tiny creatures under a microscope, but because it revealed patterns that would otherwise be impossible to detect: which microbial groups tend to travel together, how communities change after antibiotics, why two people eating similar diets can have very different microbial outputs, and why inflammation in the gut can coincide with subtle rearrangements in microbial chemistry rather than the arrival of a single obvious culprit.

Two main sequencing strategies dominate this census. The first is a roll call: *who is there?* The second is closer to reading a stack of instruction manuals: *what can they do?* Both rely on the same industrial miracle—modern DNA sequencing—but they answer different questions, with different blind spots.

A useful historical detour begins far from the gut, with a dispute about the tree of life. In the 1970s, Carl Woese was not trying to solve digestive disease; he was trying to map evolution. He focused on a gene found in all cellular life, one that changes slowly enough to preserve deep history: the small-subunit ribosomal RNA gene. In bacteria, this is the 16S rRNA gene. Woese's work helped reveal that what had been lumped together as "bacteria-like" organisms actually included a distinct domain of life: the archaea. The method was revolutionary not because it required growing every organism, but because it used a genetic fingerprint to infer relatedness. That idea—a conserved gene as an ID card—is the conceptual ancestor of today's most common microbiome survey.

16S sequencing (often called 16S rRNA gene amplicon sequencing) exploits the fact that the 16S gene contains stretches that are highly conserved across bacteria, interspersed with variable regions that differ between groups. In practice, researchers extract DNA from a stool sample, use short DNA primers to amplify (copy) a chosen region of the 16S gene via PCR (polymerase chain reaction), and then sequence the amplified fragments. A computer compares the resulting sequences to reference databases and assigns them to taxonomic categories: phylum, family, genus, sometimes species.

The appeal is obvious. It is relatively cheap. It works even when organisms refuse to grow. It can process large numbers of samples, which matters when human biology is messy and statistical power is hard-won. It answers a basic ecological question: is this community dominated by a few groups or broadly diverse? Does it shift in response to diet, illness, or medication?

Yet a roll call is not a biography. The first limitation is resolution. Many bacteria share very similar 16S sequences, particularly in the short regions typically sequenced. Two strains can behave very differently in the body—one benign, one dangerous—and still look nearly identical in a standard 16S survey. This is not a rare technical quibble; it is central to why some microbiome findings feel tantalizing but slippery. If your method cannot reliably distinguish a peaceable resident from a troublemaker cousin, you may detect a "genus-level association" without knowing which member of the family is actually present.

The second limitation is that PCR is not a neutral photocopier. Primers bind more readily to some sequences than others. Amplification favors certain templates. The number of 16S gene copies varies between bacteria, so one cell of species A might contribute more 16S fragments than one cell of species B even if their cell counts are equal. Different labs make different choices about which 16S region to amplify, which polymerase to use, how many cycles to run, and how to filter low-quality reads. Each choice can tilt the resulting census.

The third limitation is conceptual and surprisingly counterintuitive: 16S surveys (and, in a different way, many sequencing surveys) mostly give *relative* abundance, not absolute numbers. Sequencing machines produce a finite number of reads. If you sequence a sample and obtain, say, 50,000 bacterial reads, those reads must sum to 50,000. This means that when one group's proportion goes up, others must go down, even if in reality every group increased in absolute cell count. It is possible for a bacterium to appear to "decrease" in relative terms even while its absolute population rises, simply because another group expanded more dramatically. A pie chart cannot tell you whether the whole pie grew.

This matters because health often cares about the total microbial load and the chemistry it produces, not just the percentages. In diarrheal illness, for example, the gut can become a fast-flowing river that reduces total microbial biomass while changing proportions. After certain antibiotics, the overall population may crash, leaving ecological space for opportunists. A purely relative census can mislead if interpreted too literally, as if it were counting bodies rather than shares of a constrained total.

All these limitations do not make 16S useless; they make it honest about what it is. It is a fast, affordable way to detect patterns and generate hypotheses. It can tell you that something changed, and often which broad groups moved with it. It is less good at telling you precisely what changed at the strain level, and it is silent on whether the change mattered because of genes, metabolites, or sheer abundance.

16S sequencing: reading the book spines

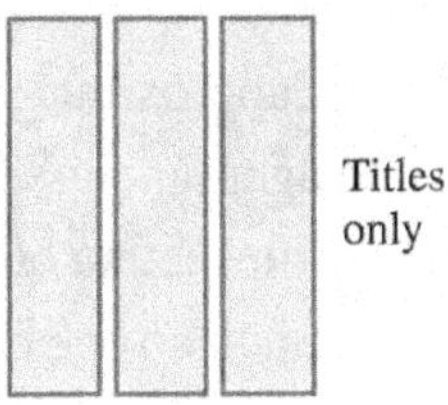

Question: Who is there?

Reads: One marker gene (16S rRNA)

Strength: Cheap, scalable, good for community shifts

Blind spots: Limited strain resolution; PCR bias; relative (compositional) data

Shotgun metagenomics: opening the books

Question: What can they do (and sometimes what are they doing)?

Reads: Many genes across genomes

Strength: Higher taxonomic and functional resolution; pathways, resistance genes

Blind spots: Cost, computation; still vulnerable to contamination; interpretation is harder

If 16S is the roll call, *shotgun metagenomic sequencing* is closer to rummaging through everyone's pockets, notebooks, and toolkits. Instead of targeting a single gene, shotgun sequencing reads fragments from all DNA in the sample: bacterial genomes, archaeal genomes, bits of viruses (including bacteriophages that infect bacteria), and often some human DNA as well. The result is not just a list of names but a partial catalogue of capabilities: enzymes for breaking down complex carbohydrates, pathways for synthesizing vitamins, genes that confer antibiotic resistance, machinery for producing inflammatory molecules, and the subtle variations that distinguish one strain from another.

The word "shotgun" is apt. You do not know in advance which fragments you will recover, only that with enough sequencing depth you will capture a representative scatter. Bioinformatics then tries to reconstruct meaning from the scatter: map fragments to known genes and genomes, estimate which organisms they came from, and sometimes assemble longer contiguous sequences that approximate whole genomes. When successful, this can yield "metagenome-assembled genomes", which sound like an administrative achievement but often feel like a revelation: a partial genome of an organism no one has ever cultured, showing a lifestyle inferred from its genes.

Function is the reason shotgun metagenomics has reshaped the field. Many of the microbiome's most important effects do not depend on the mere presence of a microbe but on what it produces. The gut is a chemical factory. Microbes turn dietary fiber into short-chain fatty acids; they modify bile acids; they transform polyphenols from plants; they generate neurotransmitter-like molecules and immune-active fragments. Two communities that look similar at the 16S level can differ in these metabolic talents, because the critical genes may be unevenly distributed across strains.

A concrete example comes from one of the most clinically urgent microbiome conditions: recurrent *Clostridioides difficile* infection. *C.*

difficile can flourish after antibiotics disrupt the normal community, causing severe diarrhea and colitis. Fecal microbiota transplantation (FMT) has shown striking success in many patients with recurrent disease, but understanding *why* it works demands more than a roll call. Shotgun approaches have helped identify patterns in functions linked to resistance against *C. difficile*, including the metabolism of bile acids. Certain gut bacteria convert primary bile acids into secondary bile acids that can inhibit *C. difficile* growth or germination. A 16S census might show that "diversity increased" after FMT, which is true but not very actionable. A functional view can point toward mechanisms: the return of specific metabolic pathways and ecological checks that keep *C. difficile* from taking over.

Shotgun sequencing has its own traps. The first is cost and complexity. Sequencing an entire metagenome requires more data than sequencing a single marker gene, especially if one wants to detect low-abundance organisms or assemble genomes. It also demands heavier computation and introduces more decisions: which reference database to use, how to handle ambiguous reads, how to account for genes shared across species, how to avoid over-interpreting fragments that could belong to multiple sources.

The second trap is that the method still often produces relative data. Yes, you may count genes rather than 16S reads, but you are still typically working within a fixed budget of sequencing depth. If the total microbial biomass changes drastically between samples, relative abundances can mislead in ways that are not immediately obvious. Researchers sometimes add external standards (known quantities of "spike-in" DNA) or pair sequencing with quantitative measures like qPCR or flow cytometry to estimate absolute load. These hybrid strategies are powerful, but they add steps, costs, and new opportunities for error. There is no free lunch in measurement.

The third trap is contamination, a problem so mundane it feels almost embarrassing, yet it has derailed serious claims. DNA extraction kits and laboratory reagents are not sterile in the genetic sense. Tiny amounts of bacterial DNA can be present, and when the sample itself contains very little microbial biomass—as in some tissue biopsies or purported "sterile" sites—contaminant DNA can dominate the signal. The result can look uncannily biological: the same contaminant taxa appear across samples, creating patterns that mimic real differences. This is why careful studies include negative controls (samples that should contain no biological DNA), track reagent lots, and use statistical approaches that flag sequences that behave like contaminants. In microbiome science, skepticism is not a personality trait; it is a laboratory technique.

Even when contamination is controlled, reproducibility can be threatened by "batch effects": systematic differences introduced when samples are processed on different days, by different technicians, with different kit lots, or sequenced on different machines. A batch effect can masquerade as biology with the confidence of a well-dressed impostor. Large multi-laboratory comparisons have shown that technical variation can sometimes rival biological variation, depending on the specimen and protocol. That does not mean results are meaningless; it means that the best microbiome studies treat method as part of the experiment, not an invisible background.

Beyond DNA lies an even more tantalizing question: not just what microbes *could* do, but what they *are doing* in a particular moment. Genes are potential. Activity is something else.

This is where the term *multi-omics* enters, a piece of jargon with an unromantic name for a very human desire: to catch the microbiome in the act. Metatranscriptomics sequences RNA to estimate which microbial genes are being expressed. Metaproteomics tries to measure proteins directly, though it is technically challenging in the crowded molecular soup of stool. Metabolomics measures small molecules—the chemical outputs that often matter most to the host. Each approach has strengths and frustrations. RNA is fragile and changes quickly. Proteins are diverse and difficult to identify without excellent databases. Metabolites can come from microbes, from diet, from human cells, or from interactions between all three, and identical molecules can be produced by different routes.

Still, when these measurements line up, they can be persuasive in a way that a simple census is not. If a person's gut community contains genes for producing a particular short-chain fatty acid, expresses those genes, and the metabolite itself rises in the stool or blood, the story becomes more than "this genus went up." It becomes a chain of evidence connecting community composition to chemistry to physiology.

There is a temptation, especially in headlines, to treat sequencing as an oracle: feed in a sample and receive truth. The reality is closer to a careful conversation with an unreliable narrator. Sequencing can reveal patterns that culture would miss entirely, but it speaks in probabilities, proxies, and partial views. It demands controls, humility, and an awareness of what the method cannot see. Even the most sophisticated metagenomics cannot, by itself, prove that a microbial change *caused* a health outcome. It can only sharpen the suspects and point toward mechanisms worth testing.

Yet for all its caveats, the genetic census has changed what it means to talk about the gut. The microbiome is no longer a vague cloud of "good" and "bad" bacteria. It is a community with identifiable members, identifiable genes, and identifiable chemical consequences. That does not make it simple. It makes it legible enough to argue with, to replicate, to falsify, and occasionally to use.

A stool sample, when reduced to DNA sequences, looks like an abstract text file. But it is also a record of ecological negotiations:

microbes competing for fiber, cooperating to digest complex carbohydrates, resisting invasion, and adapting to the host's immune landscape. Sequencing turns that negotiation into data, and data into hypotheses. The key is to remember what kind of reading you are doing. Sometimes you are scanning the spines on the shelf. Sometimes you are opening the books. Either way, the act is interpretive, and the meaning depends on how carefully you read.

9.2 Life Inside the Plastic Bubble

A sequencing plot can tell you that a certain bacterial group is more common in people with obesity, or that a particular microbial gene tends to rise before an inflammatory flare. The plot cannot tell you whether the microbe is a driver, a passenger, or merely a shadow cast by something else. Human lives are full of confounders: diets that change with stress and income, medications taken for reasons that never make it into the dataset, infections remembered imperfectly, sleep, exercise, and the small matter of genetics. Even when the statistics are careful, an uneasy question remains. If you want to claim *cause*, you need a way to intervene.

One of the boldest interventions in biology is not a drug or a surgery but an absence: an animal raised without microbes at all. Germ-free animals are not merely "very clean." They are born, delivered, and reared in conditions designed to exclude bacteria, archaea, fungi, and viruses—a life lived behind barriers so strict that the animal's food, water, bedding, and even the air it breathes are treated as potential contaminants. This is the "plastic bubble": flexible film isolators and sealed systems that look, to an outsider, like a cross between a life-support device and a child's science-fair habitat. To the scientists who work with them, it is a tool that turns suggestion into experiment.

The idea has a surprisingly long pedigree. In the late nineteenth century, Louis Pasteur speculated about whether animal life without microbes was even possible, wondering aloud if a germ-free existence might be incompatible with health. Early attempts were heroic and often grim, but by the mid-twentieth century researchers such as James Reyniers and colleagues at Notre Dame had refined the engineering and husbandry needed to keep animals free of detectable microbes. Modern gnotobiotics—the study of animals with known microbial life (from *gnōtos*, known, and *bios*, life)—is built on their legacy. The point is not sterility for its own sake. The point is control: when you start with an animal that has no microbes, you can add microbes one by one, or in defined mixtures, and ask what changes.

A germ-free mouse is often described as a "blank slate." The phrase is useful, but it can also be dangerously comforting, as if the mouse were an ordinary animal simply missing a set of tenants. In reality, the absence of microbes reshapes development from the beginning. The gut's immune tissues are under-stimulated; the normal training exercises that teach the immune system to distinguish friend from foe are altered. Structures such as Peyer's patches and the balance of immune cell types in the intestinal wall can be different from those of a conventional mouse. Even the gut itself, from motility to mucus dynamics, can behave differently. A germ-free animal is not a conventional animal minus microbes; it is an animal that has grown up under a different set of biological expectations.

That is exactly why germ-free models are so scientifically valuable. If microbes are necessary for a phenotype, the phenotype may disappear in their absence. Some classic mouse models of inflammatory bowel disease, for example, fail to develop colitis when raised germ-free. It is hard to imagine a cleaner demonstration that microbes are not just scenery in the background of inflammation; they can be part of the machinery. Once you know a disease state is microbe-dependent in this strong sense, you can begin the more delicate work of identifying which microbes, which functions, and which host pathways matter.

The most famous demonstration of "microbes as transferable traits" uses an experiment that is both conceptually simple and emotionally jarring: transplant a community of microbes from one animal into another and see whether a feature of the donor appears in the recipient. In the microbiome world this is often done using fecal microbiota transplantation, abbreviated FMT. In human medicine FMT has become a real therapy for recurrent *Clostridioides difficile* infection, but in the laboratory it is also an engine for causality.

The basic setup is this: take stool from a donor—a human, or a mouse with a particular phenotype—and introduce it into the gut of a germ-free mouse. Introduction is usually by oral gavage, a carefully performed procedure in which a small tube delivers a measured volume directly into the stomach. The germ-free mouse begins life with no microbes; after the gavage, it becomes colonized, and its gut ecosystem assembles under the influence of the donor community. If the recipient mouse begins to show a donor-like trait, the inference is powerful. Something in the microbial community can contribute to that trait.

A landmark example, discussed widely because it is so intuitive, involves obesity. Jeffrey Gordon's lab (2006) reported that germ-free mice colonized with microbiota from obese donors gained more fat than those colonized with microbiota from lean donors, even when consuming comparable diets. The result did not prove that obesity in humans is "caused by bacteria" in any simplistic way—human obesity is far too multi-factorial for that—but it did prove something more precise and more useful. A microbial community can shift energy harvest and storage in a host, and those effects can be transmitted. Suddenly, the microbiome was not merely associated with metabolism; it was a lever that could, under controlled conditions, move metabolism.

The strength of this approach lies in its clarity. The germ-free mouse begins with the same genetics, the same controlled environment, the same standardized chow. The variable introduced is the microbiota. If the phenotype changes, that is hard to dismiss. And because the system is experimental, not observational, it invites follow-up questions that human cohorts cannot easily answer. Is the effect diet-dependent? Does it require specific bacterial strains, or can different communities produce the same outcome through shared metabolic functions? Does the immune system mediate it? Does the effect persist, fade, or change across generations?

The temptation is to imagine that an FMT experiment is a tidy pipeline: donor microbes go in, donor phenotype comes out. Real life inside the plastic bubble is less tidy. Microbial communities are not static cargo; they are living collectives that compete, cooperate, and adapt. Not all donor microbes engraft equally well in a mouse. A human gut community entering a mouse intestine encounters a different anatomy, different bile acids, different mucus chemistry, and different immune pressures. Some microbes thrive, others vanish, and the resulting community can be a hybrid: shaped by the donor but remodeled by the host.

Even in mouse-to-mouse transfers, the environment is never perfectly neutral. The "cage effect" is one of the most humbling lessons in microbiome science. Mice sharing a cage share microbes through contact, bedding, and the small unglamorous realities of mouse life. Over time, cages can diverge microbiologically, even when the mice started similarly. This divergence can be strong enough to shift experimental outcomes, including inflammation measures in common models. For researchers, it means that the true unit of replication is often the cage, not the individual mouse, and that careful randomization and housing design are not bureaucratic details but biological necessities.

Then there is the daily labor of sterility itself. A gnotobiotic facility is not simply an animal room with better housekeeping. It is a choreography of protocols designed to prevent invisible intruders. Food and bedding are sterilized, often by autoclaving. Water is treated. Equipment is introduced through ports and dunk tanks. Gloves are integrated into the isolator wall so that hands never enter directly. Every transfer, every cage change, every experimental procedure is performed with the awareness that a single lapse can seed a microbial bloom that ruins months of work. The plastic film is not a metaphor. It is an engineering solution that wraps a controlled atmosphere around the animals and the tools used to care for them.

This intensity imposes trade-offs. The more sterile and sealed the system, the harder it can be to perform complex procedures. Surgery, imaging, and sophisticated behavioral assays are possible, but they require additional equipment and careful technique. The more you want the mice to experience a "normal" world, the more you compromise the very control that makes germ-free models powerful. Researchers choose among different housing systems—classic flexible film isolators, rigid isolators, sealed microisolator cages, and other contained setups—depending on whether sterility, throughput, or experimental flexibility is the priority. Each choice shapes the kinds of questions that can be asked.

What, then, counts as a convincing causal chain in microbiome science? The gold standard is rarely a single dramatic FMT result. It is a narrowing process that moves from communities to candidates. One approach begins with a microbiota transfer that reproduces a phenotype. Then, using sequencing and functional profiling, researchers look for microbes or genes that reliably track with the phenotype. Next comes reduction: simplified communities, sometimes called defined consortia, in which a germ-free mouse receives a small set of cultured strains rather than the entire donor ecosystem. Reduction is not always straightforward, because effects can depend on interactions among microbes, or on the chemical environment the community creates. Still, when it works, it is remarkably persuasive. A messy cloud of associations collapses into a testable mechanism: this strain produces this metabolite under these dietary conditions, which shifts this immune pathway, which changes this physiological outcome.

Sometimes the most convincing evidence comes not from adding microbes but from subtracting them with surgical precision. Antibiotics can be used to suppress broad groups, but they are blunt instruments and can have direct effects on the host. More targeted strategies include bacteriophages engineered or selected to attack particular bacterial strains, or genetic knockouts of specific microbial pathways in otherwise similar bacteria. These are technically demanding, but they move the field toward the kind of causal statements that medicine can use: not merely "a diverse microbiome is good," but "this function, carried out by these organisms, affects this host process."

A clinical story often sits quietly behind these laboratory maneuvers: the patient with recurrent *C. difficile*, exhausted by cycles of antibiotics and relapse. The success of FMT in many such patients has been a moral and scientific prompt. If transplanting a complex community can rescue a failing gut ecosystem, then perhaps some of the microbiome's most powerful effects are ecological rather than pharmacological. The gut may not always need a new molecule; it may need a restored set of relationships that stabilize the system and crowd out opportunists. Germ-free and gnotobiotic models do not replicate the full human experience, but they allow the ecological logic of treatments like FMT to be tested in a controlled setting: which components are necessary, which are sufficient, and which are merely correlated with recovery.

None of this removes uncertainty. Germ-free mice are extraordinary tools, but they are also extreme organisms. A phenotype that appears in a germ-free recipient after colonization might depend on the recipient's unusual baseline physiology. A microbe that drives inflammation in a mouse might behave differently in a human gut shaped by decades of diet, infection, and immune memory. Even within the laboratory, small environmental differences can steer microbial assembly in different directions. The lesson is not cynicism; it is precision. Germ-free models can demonstrate that microbes *can* cause a change, and they can help identify plausible mechanisms. They cannot, by themselves, guarantee that the same mechanism dominates in every human body.

A useful way to hold the power and the limits in mind is to picture what the plastic bubble truly offers: not a miniature human, but a controllable ecosystem. Inside that ecosystem, a community of microbes becomes an experimental variable rather than an afterthought. That shift—from measuring associations in a crowded human world to manipulating communities in a controlled animal world—is one of the reasons the microbiome has become experimentally tractable at all.

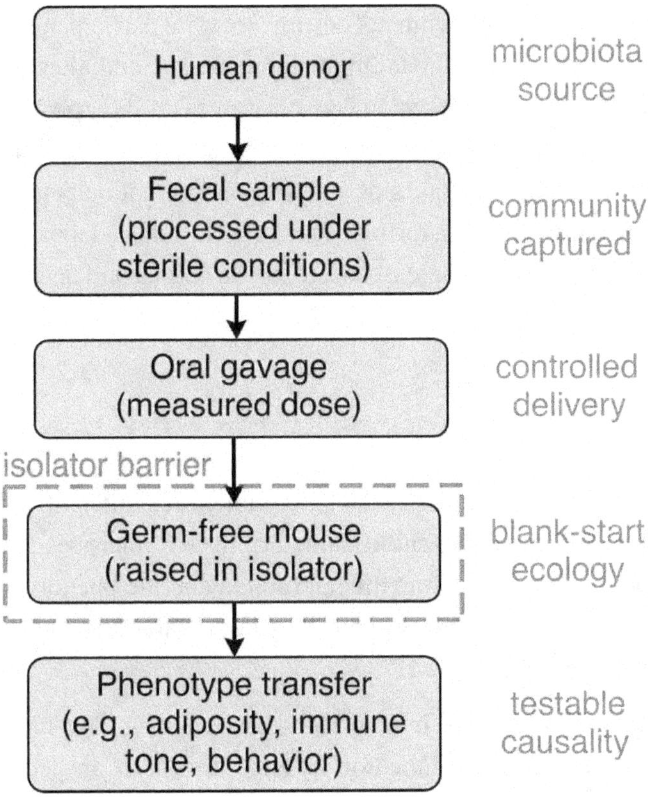

There is a moment in many gnotobiotic experiments that carries an almost theatrical charge. The mice have been raised in microbial silence. Then, with a few drops of liquid delivered by gavage, an ecosystem arrives. Within days, the gut changes its chemistry. The immune system begins to receive signals it has never received. The animal's physiology adjusts, sometimes subtly, sometimes dramatically. The timeline compresses what, in a human, is a lifelong accumulation of exposures. It is hard not to feel, in that moment, that biology is being rewritten by something that is neither purely genetic nor purely environmental, but a living interface between the two.

The most responsible reading of these experiments is also the most interesting. Microbes are not destiny; they are participants. They can tilt metabolism, modulate immune thresholds, and alter susceptibility to disease, but always in conversation with diet, host genes, and the broader environment. Germ-free models make that conversation audible by silencing one side and then restoring it on command. The plastic bubble is less a fortress against germs than a laboratory instrument for testing ecological causation, one animal and one community at a time.

9.3 The Rodent-to-Human Gap

A germ-free mouse can make an argument feel airtight. Give it microbes from donor A and it gains fat; give it microbes from donor B and it stays lean. Swap the microbiota and the phenotype follows. Compared with the messy sprawl of human life, the logic is almost elegant enough to frame.

Then, too often, the human trial disappoints. The microbe that seemed to calm inflammation in mice does little in patients. The probiotic that brightened mouse behavior fails to outperform placebo. A microbial signature that looked like a beacon in animals becomes a faint smudge in people, hard to separate from diet, medications, and the background noise of ordinary variation. This is not because the mouse experiments were fraudulent, or because human biology is perversely uncooperative. It is because the distance between rodent and human is real, and the microbiome amplifies that distance in ways that can be subtle until they suddenly aren't.

The first gap is the most obvious one: mice are not small humans. Their intestines differ in size, shape, and flow. Their diet in the lab is often a standardized chow that no human would willingly eat for long: a compressed, shelf-stable mix with a specific fiber profile, fat content, and micronutrient composition. Their daily rhythms, stress responses, and social environments are profoundly different. Their immune systems, while sharing many components with ours, are tuned differently at baseline. Even the mucus layer that coats the gut lining—the habitat in which many microbes live—is not identical in composition or structure across species. If microbes are sensitive to habitat (and they are), then moving a human community into a mouse gut is less like moving a potted plant from one windowsill to another and more like transplanting it across climates.

A second gap sits inside the microbiome itself. The community is not just a bag of species; it is a network of interactions shaped by long cohabitation with a host. Some human gut microbes appear to be specialists in the human intestine, adapted to our bile acids, our immune signals, and our dietary patterns. When they enter a mouse, many do not establish well. Others establish but behave differently, because the neighbors they expect are missing or because key nutrients arrive in different forms. Human microbiota-associated rodents often end up with an ecosystem that resembles the donor in broad strokes but diverges in important details. The mouse becomes a translator who speaks with an accent, and sometimes changes the meaning of the sentence.

Engraftment is not a technical nuisance; it can change conclusions. Suppose a particular human bacterium is rare but potent, producing a metabolite that shapes immune tone. If that bacterium fails to colonize the mouse, the mouse experiment might suggest the donor microbiota has no effect, when in fact the relevant component simply never took hold. Conversely, a bacterium that blooms in the mouse may look like a culprit even if it is a minor player in humans. Without checking which taxa and functions actually engrafted, it is easy to mistake a failed transfer for negative evidence, or a mouse-specific bloom for a human-relevant mechanism.

A third gap arises from the way laboratory mice are made consistent.

Consistency is the point: the promise of animal models is that they reduce variation so that cause and effect can be seen. But the microbiome is not merely a variable to be controlled; it is an ecosystem that responds to control itself. Laboratory mice have their own characteristic microbiotas shaped by vendor, facility, and husbandry. Mice ordered from different suppliers can arrive with distinct microbial starting points, and those differences can persist even after weeks in the same room. Diet, bedding, cage cleaning schedules, water treatment, and antibiotic exposures all shape the microbial baseline. Two labs can run "the same" experiment and obtain different results because their mice begin with different microbial contexts that interact with the intervention.

This is one reason microbiome papers sometimes read like detective stories. A lab observes that a genetic knockout mouse develops a striking immune phenotype. Another lab, using ostensibly the same knockout line, sees a weaker phenotype or none at all. The culprit is not always a mislabeled mouse or sloppy technique. Sometimes it is a difference in microbes: one facility happens to harbor a bacterium that triggers inflammation in susceptible hosts, while another facility does not. The gene appears causal in one context and irrelevant in another because the environment that reveals its importance is microbial. In conventional genetics, background strain effects are a known complication. In microbiome-influenced phenotypes, background microbial effects can be even more dramatic.

The cage effect adds a further twist. Mice do not have independent microbiomes when they share space. They share microbes like roommates sharing colds, except that the sharing is constant and the 'cold' is the entire ecosystem. Over time, cages drift into distinct microbial neighborhoods. If an experiment treats each mouse as an independent data point while ignoring cages, it can accidentally count the same microbiological reality multiple times. A treatment may appear to have an effect simply because one group happened to be housed in cages that drifted in a particular direction. The more subtle version of this error is that cages do not just confound; they can interact with interventions. A dietary change may have a strong effect in one cage and a weak effect in another because the starting microbes differ. In that situation, averaging across mice can hide the biology that matters most: some communities respond, others resist.

Maternal transmission is the earliest cage effect. A mouse's first microbiome is shaped by its mother and littermates. If experimental groups are divided in a way that clusters siblings together, apparent treatment effects can actually reflect inherited microbial differences. This is why careful experimental design in microbiome studies often involves splitting litters, co-housing, or cross-fostering, depending on the question. These are not fussy details; they are attempts to ensure that an experiment is about the intended variable rather than the invisible family tree of microbes.

A historical cautionary tale comes from outside microbiome research but captures the broader point: animal models can be precise and still mislead if their precision is aimed at the wrong target. In the early twentieth century, researchers made elegant dietary deficiency models in animals, producing consistent symptoms with consistent interventions. Translating those models to humans sometimes worked brilliantly (as with scurvy and vitamin C), and sometimes stumbled because human diets, genetics, and exposures were more varied than the model assumed. The microbiome adds a similar layer of hidden variability. A mouse model can yield a clean mechanistic pathway that is real in mice, yet represent only one of several possible pathways in humans, or a pathway that matters only in a subset of people whose microbial ecologies match the model's assumptions.

The immune gap deserves special attention because so many microbiome claims, implicitly or explicitly, hinge on immune modulation. Mice and humans share many immune cell types and signaling molecules, but their baseline immune landscapes differ. The frequency of certain cells, the balance of innate effectors, the anatomy of gut-associated lymphoid tissue, and the lifetime of antigen exposures are not the same. Laboratory mice also tend to live in comparatively sheltered microbial worlds. A wild mouse, or a pet-store mouse, carries a history of infections and environmental exposures that can "mature" the immune system in ways that make it behave more like an adult human's. This matters because the immune system is not a static target. A microbial intervention that nudges an immature immune system toward balance may have a different effect in an immune system already shaped by decades of viral encounters, allergies, vaccinations, and chronic low-grade inflammation.

Diet is another translator that often changes the meaning of microbiome results. Microbes live on the leftovers: fibers humans cannot digest, mucus secretions, bile acids, and compounds produced by other microbes. Changing diet changes which leftovers arrive. Many mouse studies use diets that differ dramatically in fiber content, fat type, emulsifiers, or micronutrients compared with human diets. A bacterium that produces beneficial metabolites on a fiber-rich diet may produce very little on a low-fiber diet. A community that resists invasion when fiber is plentiful may become fragile when the diet is stripped down. If a therapy works in mice fed a particular chow, it may not work in humans whose diets provide a different substrate landscape.

Even the physical architecture of the gut matters. The mouse cecum is proportionally large and functions as a major fermentation chamber. In humans, fermentation is distributed differently, and transit times differ. These anatomical differences shape gradients of oxygen, pH, and nutrients along the gut, which in turn shape microbial niches. When a human microbe "chooses" to live in a particular part of the colon, it is responding to a set of physical constraints that may not be replicated in a mouse.

These gaps can sound discouraging, but they can be turned into a practical skepticism toolkit: a way to read microbiome animal studies with curiosity rather than credulity.

One question is always: *what exactly was transferred?* If a paper claims a phenotype is transmissible by FMT, the critical follow-up is whether the investigators measured engraftment and in what way. Did they show that donor taxa appeared in recipients? Did they track not just names but functions, such as metabolic pathways or metabolites? A phenotype might transfer even if the taxonomic profile shifts, because different microbes can sometimes perform similar functions. Conversely, a failure to replicate might reflect poor engraftment rather than a false hypothesis.

Another question is: *what is the unit of replication?* If treatment is applied to mice but cages are not balanced, the data may look more definitive than they are. Well-designed studies often treat cage as a central variable, randomize at the cage level, or at least analyze with the cage structure in mind. When that is absent, strong claims deserve a raised eyebrow.

A third question is: *what is the baseline microbiome, and how stable is it?* Studies that report vendor, diet composition, housing conditions, and microbial monitoring provide the context needed to interpret results. Without this, the experiment can be hard to situate. A striking effect in one facility might be a facility-specific interaction rather than a generalizable mechanism.

A fourth question concerns *directionality*. Because sequencing commonly yields relative abundance, an apparent "increase" in a taxon could reflect a decrease in others. If the study pairs sequencing with absolute quantification, or measures microbial load, the interpretation becomes more robust. In animal work, this is often feasible and can strengthen causal inference.

Then comes the question that matters most for translation: *does the experiment model a realistic human intervention?* A germ-free mouse is a powerful causal tool, but it is not a common human state. Antibiotic-treated mice are closer to certain clinical contexts, but antibiotics have their own host effects. Defined consortia can be closer to a therapeutic product, but simplified communities can behave differently from complex ones. When a study uses an extreme setup, the value lies in mechanistic insight; the challenge lies in knowing how much of that mechanism survives in the more crowded human world.

Concrete examples help illustrate how these questions play out. Consider the idea that specific microbes can shape "anxiety-like behavior" in mice. There is intriguing animal evidence that microbial changes can influence stress responses, sometimes through immune signaling, sometimes through microbial metabolites that affect the nervous system, sometimes through vagus nerve pathways. Yet translating "anxiety-like behavior" from mouse assays to human mental states is inherently fraught. Mouse behavior tests often measure exploration, avoidance, or immobility in specific setups; these can be influenced by many factors besides anxiety. If a microbial intervention changes activity levels, metabolism, or inflammation, it can shift behavior scores without mapping neatly onto human anxiety. The result may still be biologically meaningful—the gut can influence the brain—but the human claim must be framed with restraint. The mouse can tell you that microbes can influence neural-relevant pathways; it cannot tell you that a yogurt drink treats generalized anxiety disorder.

By contrast, the translation pathway for recurrent *C. difficile* infection has been more straightforward because the outcome is concrete:

recurrence of a specific infection. Even here, though, animal models simplify. In humans, antibiotics, hospital exposures, age, immune status, and co-morbidities interact with microbial ecology. FMT can work spectacularly, but it is not a universal reset button. Regulatory agencies have also had to grapple with safety: the stool community is a complex biological product that can carry pathogens or resistance genes if screening fails. Mice can help identify mechanisms and candidate consortia, but they cannot fully model the safety and variability landscape of human donors and patients.

The most reliable bridge from rodent causality to human relevance is often a convergence of evidence. A microbial pathway discovered in mice becomes more compelling when a related signature appears in human data, especially when it is linked to measurable chemistry in blood or stool. Human trials become more interpretable when they incorporate mechanistic readouts: did the candidate microbe colonize, did the metabolite change, did an immune marker shift in the predicted direction? Without these checks, a negative trial can be ambiguous. With them, it can be informative: the intervention might have failed because it did not engraft, because diet prevented its activity, or because the pathway matters only in certain microbial contexts.

There is also a cultural temptation to treat "animal evidence" as automatically more causal than "human evidence." In microbiome science, that hierarchy can be misleading. A meticulously designed human randomized controlled trial, even if modest in size, can provide a kind of causality that no mouse can match for human outcomes. The difficulty is that microbiome interventions in humans are hard to standardize: diet varies, adherence varies, baseline microbiomes vary, and the placebo effect is powerful for subjective outcomes. Still, when a human trial is well controlled and includes microbial and metabolic measurements, it can do something animal models cannot:

it can tell you whether changing the microbiome in a particular way actually changes health in the species that matters.

A final, sobering point is that microbes are not single agents in the way a drug molecule is. They evolve, they exchange genes, and they respond to context. A strain that behaves one way in one host can behave differently in another. A community that is stable under one diet can become unstable under another. This is not a flaw in the science; it is the nature of ecosystems. The mistake is to promise simple, universal rules where the underlying biology is conditional.

The rodent-to-human gap does not invalidate animal models. It sharpens the kind of claims they can support. A well-executed mouse experiment can establish that a microbe or microbial function *can* drive a host change, identify plausible mechanisms, and reveal which host pathways are susceptible to microbial influence. Translation requires a second step: showing that the same mechanism operates, at least sometimes, in the more diverse, historically burdened, and environmentally complex human gut.

A good microbiome claim, by the time it reaches the public, ought to carry some modesty in its language and some specificity in its mechanism. Instead of "gut bacteria cause disease X," the more honest sentence is usually longer: under certain diets, in certain microbial contexts, a set of microbial functions can shift host physiology in ways that raise or lower risk. That longer sentence is harder to sell, but it is closer to how bodies actually work. When medicine eventually harnesses the microbiome with the reliability people hope for, it will likely look less like a universal probiotic and more like an ecological prescription: matching microbes, diet, and host context with the care one would bring to any complex organ system.

www.ingramcontent.com/pod-product-compliance
Lightning Source LLC
LaVergne TN
LVHW020234210926
848481LV00016B/1856